SpringerBriefs in Earth System Sci

Series Editors
Kevin Hamilton
Gerrit Lohmann
Lawrence A. Mysak

For further volumes:
http://www.springer.com/series/10032

Rupert Ford · Graham Riley
Reinhard Budich · René Redler

Earth System Modelling – Volume 5

Tools for Configuring, Building
and Running Models

 Springer

Rupert Ford
School of Computer Science
The University of Manchester
Oxford Road, Manchester
M13 9PL, UK
e-mail: rupert@manchester.ac.uk

Graham Riley
School of Computer Science
The University of Manchester
Oxford Road, Manchester
M13 9PL, UK
e-mail: graham.riley@manchester.ac.uk

Dr. Reinhard Budich
Max-Planck-Institut für Meteorologie
Bundesstraße 53
20146 Hamburg, Germany
e-mail: reinhard.budich@zmaw.de

Dr. René Redler
Max-Planck-Institut für Meteorologie
Bundesstraße 53
20146 Hamburg, Germany
e-mail: rene.redler@zmaw.de

ISSN 2191-589X
ISBN 978-3-642-23931-1
DOI 10.1007/978-3-642-23932-8
Springer Heidelberg Dordrecht London New York

e-ISSN 2191-5903
e-ISBN 978-3-642-23932-8

Printed on acid-free paper

Springer is part of Springer Science+Business Media (www.springer.com)

Preface

Climate modelling in former times mostly covered the physical processes in the atmosphere. Nowadays, there is a general agreement that not only physical, but also chemical, biological and, in the near future, economical and sociological—the so-called anthropogenic—processes have to be taken into account on the way towards comprehensive Earth system models. Furthermore these models include the oceans, the land surfaces and, so far to a lesser extent, the Earth's mantle. Between all these components feedback processes have to be described and simulated.

Today, a hierarchy of models exist for Earth system modelling. The spectrum reaches from conceptual models—back of the envelope calculations—over box-, process- or column-models, further to Earth system models of intermediate complexity and finally to comprehensive global circulation models of high resolution in space and time. Since the underlying mathematical equations in most cases do not have an analytical solution, they have to be solved numerically. This is only possible by applying sophisticated software tools, which increase in complexity from the simple to the more comprehensive models.

With this series of briefs on "Earth System Modelling" at hand we focus on Earth system models of high complexity. These models need to be designed, assembled, executed, evaluated and described, both in the processes that they depict as well as in the results the experiments carried out with them produce. These models are conceptually assembled in a hierarchy of submodels, where process models are linked together to form one component of the Earth system (Atmosphere, Ocean, ...), and these components are then coupled together to Earth system models in different levels of completeness. The software packages of many process models comprise a few to many thousand lines of code, which results in a high complexity of the task to develop, optimise, maintain and apply these packages, when assembled to more or less complete Earth system models.

Running these models is an expensive business. Due to their complexity and the requirements w.r.t. the ratios of resolution versus extent in time and space, most of these models can only be executed on high performance computers, commonly called supercomputers. Even on todays supercomputers typical model experiments take months to conclude. This makes it highly attractive to increase the efficiency of the codes. On the other hand the lifetime of the codes exceeds the typical lifetime of computing systems, including the specifics of their architecture, roughly by a factor of 3. This means that the codes need not only be portable, but also constantly adapted to emerging computing technology. While in former times computing power of single processors—and consequently of clustered computers—was resulting mainly from increasing clock speeds of the CPUs, todays increases are only exploitable when the application programmer can make best use of the increasing parallelism off-core, on-core and in threads per core. This adds additional complexity to areas such as IO performance, communication between cores or load balancing to the assignment at hand.

All these requirements place high demands on the programmers to apply software development techniques to the code making it readable, flexible, well structured, portable and reusable, but most of all capable in terms of performance. Fortunately these requirements match very well an observation from many research centres: due to the typical structure of the staff of the research centres, code development often has to be done by scientific experts, who typically are not computing or software development experts. So code they deliver deserves a certain quality control to assure fulfilment of the requirements mentioned above. This quality assurance has to be carried out by staff with profound knowledge and experience in scientific software development and a mixed background from computing and science.

Since such experts are rare, an approach to ensure high code quality is the introduction of common software infrastructures or frameworks. These entities attempt to deal with the problem by providing certain standards in terms of coding and interfaces, data formats and source management structures etc., that enable the code developers as much as the experimenters to deal with their Earth system models in a well-acquainted and efficient way. The frameworks foster the exchange of codes between research institutions, the model inter-comparison projects so valuable for model development and the flexibility of the scientists when moving from one institution to another, which is commonplace behaviour these days.

With an increasing awareness about the complexity of these various aspects scientific programming has emerged as a rather new discipline in the field of Earth system modelling. Coevally new journals are being launched providing platforms to exchange new ideas and concepts in this field. Up to now we are not aware of any text book addressing this field, tailored to the specific problems the researcher is confronted with. To start a first initiative in this direction, we have compiled a series of six volumes each dedicated to a specific topic the researcher is confronted with when approaching Earth system modelling:

Volume 1—Recent Developments and Projects
Volume 2—Algorithms, Code Infrastructure and Optimisation
Volume 3—Coupling Software and Strategies
Volume 4—IO and Postprocessing

Volume 5—Tools for configuring, building and running models
Volume 6—ESM Data Archives in the Times of the Grid

This series aims at bridging the gap between IT solutions and Earth system science. The topics covered provide insight into state-of-the-art software solutions and in particular address coupling software and strategies in regional and global models, coupling infrastructure and data management, strategies and tools for pre- and post-processing and techniques to improve the model performance.

Volume 1 familiarizes the reader with the general frameworks and different approaches for assembling Earth system models. Volume 2 highlights major aspects on design issues that are related to the software development, its maintenance and performance. Volume 3 describes different technical attempts from the software point of view to solve the coupled problem. Once the coupled model is running data are produced and postprocessed (Volume 4). Volume 5 at hand addresses the whole process of configuring a coupled model, running the model and processing the output. Compared to the focus in the previous volumes the reader is made familiar with software tools and concepts that help building up the related workflow and ease the handling of the complex task chain. Lastly, Volume 6 describes coordinated approaches to archive and retrieve data.

Hamburg, June 2011   Reinhard Budich
                                                                                 René Redler

Acknowledgments

The Editors and Authors would like to thank Brian Eaton, John Michalakes and Nigel Wood for their invaluable reviews which greatly improved the quality of this book.

Contents

Contributors

V. Balaji Princeton University, Princeton, USA, e-mail: balaji@princeton.edu

Mick Carter Met Office Hadley Centre, Exeter, UK, e-mail: mick.carter@metoffice.gov.uk

Rupert W. Ford School of Computer Science, University of Manchester, Manchester, UK, e-mail: rupert@manchester.ac.uk

Sudipta Goswami Tyndall Centre, University of East Anglia, Norwich, UK, e-mail: s.goswami@uea.ac.uk

Amy Langenhorst High Performance Technology, Inc., USA, e-mail: amy.langenhorst@noaa.gov

Claes Larsson UReason, Maidenhead, UK, e-mail: clarsson@ureason.com

Stephanie Legutke German Climate Computing Centre (DKRZ), Hamburg, Germany, e-mail: legutke@dkrz.de

David Matthews, Met Office, Exeter, UK, e-mail: david.matthews@metoffice.gov.uk

Graham D. Riley School of Computer Science, University of Manchester, Manchester, UK, e-mail: graham.riley@manchester.ac.uk

Ufuk. U. Turuncoglu Istanbul Technical University, Istanbul, Turkey, e-mail: u.utku.turuncoglu@be.itu.edu.tr

Rachel Warren, Tyndall Centre, University of East Anglia, Norwich, UK, e-mail: r.warren@uea.ac.uk

Gethin Williams School of Geographical Sciences, University of Bristol, Bristol, UK, e-mail: gethin.williams@bristol.ac.uk

Chapter 1
Introduction

Rupert Ford and Graham Riley

This book is the fifth volume in a series of books on Earth system modelling (ESM) infrastructure[1] and builds on the content of the earlier volumes.

The book begins at the point where the issues of the scientific contents of models have reached a level of maturity and codes have been developed (see Vol. 2), coupling technologies and techniques exist (see Vol. 3), and mechanisms for getting data into and out of models (i.e. I/O and archiving systems) are available (see Vol. 4). Now, in the context of a process of continuous improvement and development of scientific codes and software infrastructures, scientists are in a position to construct and run computational experiments to investigate specific scientific questions.

This book is concerned with how to construct and configure (coupled) models, build the appropriate executable or executables from the relevant mix of source code and libraries, specify and access any input data, and finally, run the executable, or executables, either as a single *job* or, for example, in a statistical *ensemble* or parameter search/tuning *experiment*. In this book any individual processes associated with the above activities are classed as *workflow processes* and the overall coordination of these activities and their presentation to the user is collectively classed as *workflow management*.

The major workflow processes that have to be managed include *source code version control*—in the context of a dynamic software development environment, *configuration management*, which includes coupled model construction, *building* executable(s) from source code and/or libraries, *execution control*, i.e. the staging and

---

[1] Please see preface for more details.

R. Ford (✉) · G. Riley
School of Computer Science, The University of Manchester,
Oxford Road, Manchester, M13 9PL, UK
e-mail: rupert@manchester.ac.uk

G. Riley
e-mail: graham.riley@manchester.ac.uk

running of models, and the *monitoring* of active jobs and experiments, monitoring both the scientific progress of the experiments and the progress of the programs involved as they execute on the computing systems used. The *Post processing* and *archiving* of data, whilst important workflow processes, are only touched on in this book as they are discussed in Vols. 4 and 6 of this series.

The task of workflow management is relatively straightforward when there is a single user who is also the code developer and the model(s) that s/he is developing are relatively small in terms of software complexity (measured, for example, in terms of the number of source files and the number of lines-of-code). Here, simple scripting might be used to select appropriate source and configuration files (specifying, amongst other things, input files and destinations for output etc.), manage the compilation of the source for the target machine (perhaps a desktop machine but possibly a large, parallel supercomputer or even a Computational Grid or the Cloud), and submit the resulting job(s) for execution (perhaps via an appropriate batch queueing system).

For scientists operating in a larger group or institution, more elaborate workflow processes and management is required as model code is usually much more complex, typically being developed collaboratively by different groups of scientists, used in many different configurations and potentially shared between institutions. This, more complex, workflow is the primary focus of this book.

The book is organised into seven chapters[2] each of which has been written by domain experts from, or associated with, major research groups in Earth system modelling and Integrated Assessment. Whilst the chapter content is primarily from the perspective of the centre where the domain experts work (or previously worked), the issues raised are representative of the community. Any associated tools or techniques described are state-of-the-art but that is not to say that there are not other valid approaches in other institutions and indeed one of the contributions of the book is to point out the limitations of current solutions and relate them to others, where appropriate.

The first two chapters discuss the topic of workflow management from the perspective of complex ESM systems, as are typically run at the major modelling centres. These chapters address the problem of how to bring workflow tasks together to enable the user to construct and run individual jobs or experiments (or sets of jobs in the case of ensemble experiments etc.) and manage the resulting output.

In the first of these two chapters, Balaji and Langenhorst give an overview of the salient issues in workflow management and present an ESM-specific workflow solution (FRE). They talk about how ESM is transitioning from being a specialist task to a general purpose tool and propose that standard, generic ESM workflows can help this process.

In the second of the two workflow chapters Turuncoglu presents the application of a generic workflow system (Kepler) to ESM. He illustrates the issues by using CCSM4 as a motivating example and demonstrates that Kepler could be relatively easily specialised to support this particular example.

---

[2]   There is also a conclusions chapter and this introductory chapter.

In the following three chapters each of the workflow processes described earlier (*source code version control and configuration management, building* executable(s) from source code and/or libraries and *execution control and monitoring*) are discussed in more detail, again from the perspective of the major modelling centres. Each of these chapters describe the requirements of a particular workflow process and compare these requirements with the current state of the art, with use case[3] walk-throughs of existing systems being provided, where appropriate. The chapters conclude with a look at future trends and a summary.

In the first of the three workflow process chapters (Chap. 4), Carter and Matthews provide insight into the issues that a large ESM organisation must face when developing and constructing ESM models. They identify and discuss a number of distinct roles within such an organisation and outline tools currently in use to support these roles. They identify reproducibility as an important issue in configuration management and point to the trend towards increased collaboration between different institutions as representing a challenge for configuration management.

In the second of the three workflow process chapters (Chap. 5) Legutke presents the requirements of an ESM build system and identifies a number of key issues including build times and portability of software between different architectures. She introduces and discusses techniques and tools that are used by ESM centres to help meet their requirements and illustrates the issues using SCE as a use case.

In the third of the three workflow process chapters (Chap. 6), Larsson introduces job monitoring and management. He outlines a number of tools that can be used for job monitoring and management and presents a detailed use case using SMS. He concludes with a vision of future monitoring and management systems supporting jobs running on remote Cloud or Grid systems and allowing submission and monitoring from any type of networked device at any time of the day.

Up to this point the book discusses workflow management and workflow processes from the perspective of the major modelling centres, as this is the primary aim of the series. However, it is also informative to examine domains closely related to ESM as these may provide insight into the problems that ESM is currently facing and/or may have to face in the future. One particularly convincing argument for looking at Integrated Assessment in particular is the observation that "the emphasis for ESM is shifting from global modeling that established anthropogenic climate change to regional modeling for assessing impacts, adaptation, and remediation. The computational and software requirements will necessarily be different in this new regime."[4]

Therefore, the following two chapters present case studies of workflow management and processes from a paleo-climate model (which is effectively ESM on a longer time-scale) and an Integrated Assessment model (which effectively integrates

---

[3]   A use case is a methodology used in system analysis to identify, clarify, and organize system requirements. The use case is made up of a set of possible sequences of interactions between systems and users in a particular environment and related to a particular goal (as defined by http://searchsoftwarequality.techtarget.com).

[4]   John Michalakes (http://www.mmm.ucar.edu/people/michalakes) in a response to a draft of this book.

ESM into Integrated Assessment on a finer grid scale). Each chapter highlights any similarities and differences in the requirements and proposed solutions with those from the major modelling centres.

The first of these two chapters (Chap. 7) presents workflow management and processes from the perspective of a community model (GENIE) designed for long-scale and paleo-climate simulations. Many of the issues and solutions employed in this model are the same as those described in earlier chapters. However there are also interesting differences. For example, in the GENIE community model, portability and performance across different architectures and compilers is considered to be very important and this is reflected in their approach.

The second of these two chapters (Chap. 8) presents workflow management and processes from the perspective of an Integrated Assessment modelling system (CIAS). Again, a number of the issues and solutions employed in this model are the same as those described in earlier chapters. However a notable difference is that Integrated Assessment demands a much greater focus on interoperability between components, as models are written in different institutions and in different computer languages.

Finally, the last chapter (Chap. 9) concludes the book with a summary of the main points that have emerged from the earlier chapters.

Chapter 2
ESM Workflow

V. Balaji and Amy Langenhorst

The period 2000–2010 may be considered the decade when Earth system modeling (ESM) came of age. Systematic ESM-based international scientific campaigns, such as the Intergovernmental Panel of Climate Change (IPCC), are recognized as central elements both in scientific research to understand the workings of the climate system, as well as to provide reasoned and fact-based guidance to global governance systems, on how to deal with the planetary scale challenge of climate change.

The running and analysis of ESMs is a huge technical challenge, whose several elements have been discussed in previous volumes. The problems range from assembling models from across a community of specialists in different aspects of the climate—oceanographers and stratospheric chemists, for example—to executing climate models on complex supercomputing hardware, to developing global distributed data archives for coordinated international experiments.

At the modeling centers, we are experiencing a serious escalation in the demands placed on us. There is an expectation that ESMs will become "operational": used as forecast tools on time scales of seasons to decades; used to generate scenarios of future climate change in response to particular policy choices; used as a planning tool in industries from energy to insurance.

This trend implies a radical shift in the way we do modeling: *an integrated infrastructure for the building and running of ESMs and analysis of ESM output data.* What is now a heroic effort involving a motley crew of scientists, software engineers and systems analysts has to become a streamlined process: a *scientific workflow.*

V. Balaji (✉)
Princeton University, Princeton, USA
e-mail: balaji@princeton.edu

Amy Langenhorst
High Performance Technology Inc, Princeton, USA
e-mail: amy.langenhorst@noaa.gov

R. Ford et al., *Earth System Modelling – Volume 5,* SpringerBriefs in Earth System Sciences, DOI: 10.1007/978-3-642-23932-8_2, © The Author(s) 2012

2.1 What is a Workflow?

A *workflow* is a more or less formal description of a sequence of activities. The description is intended to be precise and comprehensive. In the context of science, it provides *reproducibility* of an experiment. For sciences such as Earth system modeling that are based on digital simulation, this can in principle allow *exact binary reproducibility* of an experiment.

Once a workflow is fully described, it can not only be reproduced exactly; it then becomes easy to define perturbations on those experiments as well, for instance by rerunning the experiment varying a single parameter. Two such perturbed experiments can be compared by differencing the workflows: every element in the workflow will be identical save the one parameter that was varied.

Finally, the formal description serves as input to a *curator* of model output. Since all of the scientific, numerical and technical input into the model is specified in the workflow, it serves as a complete descriptor of the model output as well.

In the next section we provide an example of a functioning workflow system, and use it develop a general architecture for workflows and curators in the context of Earth system modeling.

2.2 Example: FRE, the FMS Runtime Environment

Weather centres and operational sites have incorporated workflows for running their models [e.g. prepIFS developed at European Centre for Medium-Range Weather Forecasts (ECMWF) and in use at several weather centres Larsson and Wedi (2006)] but those tend to have restricted palettes of choices, more attuned to operational uses. In this section, we describe in some detail a scientific workflow for ESMs aimed at development and testing as well as operational running of models, as an example to illustrate the range of issues one must deal with in workflow development. FRE, the FMS Runtime Environment is a workflow system developed for the Flexible Modeling System (FMS) described in Volume 3 of this series. FRE has been an operational system for models at the Geophysical Fluid Dynamics Laboratory (GFDL) since 2002, and was a critical element in the process of running GFDL models for IPCC AR4 in 2003–2004, and ever since.

FRE was developed with the goal of simplifying the process of assembling, running and analyzing FMS-based models. Steps in the modeling process include:

- *retrieve known site configuration information.* For instance, best-practice compiler and library configurations, paths to needed standard and bespoke utilities for data processing; instructions to a batch-processing system.
- *retrieve known component configurations in a multi-component model.* As in many ESMs that follow component-based design, an FMS model is composed of many components. Within FRE, an oceanographer would be able to focus on maintaining exquisite control of her own component, while inheriting an atmosphere component's configuration wholesale from another experiment.

- *perform basic tests before launching a run.* FMS best practice requires running some standard tests before launching. A basic compatibility check tests that the ESM can be run forward stably and safely for short duration. Further, since we know that an ESM run involves a long series of runs, we must verify that the model saves state ("restart files") with adequate information. A simple test of this is to run the model for 2 days in a single run, and then again in two segments of 1 day. At the end, both runs should provide bitwise-identical results. A second test might be to see if the model provides the same answer at different processor counts. Another form of basic test might be to check the integrity of input data files which might have been acquired from a remote repository, by verifying their checksums.
- *launch, manage and monitor a long-running job sequence.* A single job from an ESM simulation submitted to a batch system is usually a small fraction of the entire run length. FRE jobs are designed to resubmit themselves until a specified end time, or to run indefinitely until manually interrupted. A user is able to query the model state.
- *launch, manage and monitor post-processing , analysis and data publication.* Post-processing (see Vol. 4) is itself an intensive task rivalling in resource consumption the model run itself. This step can take us all the way to the preparing of graphical output for a scientific audience, and publication of data on a public portal.

An experimental configuration is expressed in an XML (Bray et al. 2000) file called a "FRE file". The Extensible Markup Language (XML) is a common choice for such applications for various reasons: it is textual, easy to manipulate and edit, with a rich and readily-available environment of tools and software.

A single FRE file contains the complete workflow of an FMS experiment. Thus at any stage, any aspect of the experiment may be queried: a user looking at graphical output from an experiment might ask for the value of some input parameter, or what tracer advection scheme was used by the ocean, or how many processors were used by the job; all of this information is at her fingertips, in the FRE file.

The various stages in the workflow are each executed by a script, which may be interactive on the desktop, or remotely executed on a batch system. The scripts generated by FRE are generally in the C-shell (csh). These shellscripts themselves are generated by command-line[1] "meta-scripts" that process and interpret the FRE file to generate the requisite instructions: for instance fremake generates compilation instructions for the experiment; frerun generates the runscript to execute the experiment. These "meta-scripts", known as FRE scripts, are written in languages like perl and python.

A fragment of a FRE file is shown here for discussion:

The XML fragment above illustrates some key aspects of FRE workflow.

- The basic workflow unit of FRE is an <experiment>.

---

[1] FRE at the moment of writing is entirely command-line based: a graphical user interface is conceivable, but has yet to be built. We note that different labs have cultural biases toward the command line or the GUI: GFDL appears to be of the textual rather than iconographic culture.

```
<experiment name="CM2.4C_U1" inherit="CM2.4C_base">
    <component name="fms">
        <source vc="cvs" root="/home/fms/cvs">
            <codeBase version="perth">shared</codeBase>
            <csh>
                cvs up -r perth_bw time_interp.F90
            </csh>
        </source>
        <compile>
            <cppDefs>-Duse_libMPI -Duse_netCDF</cppDefs>
        </compile>
    </component>
    <component name="mom4p1" requires="fms">
        <compile>
            <cppDefs>"-DUSE_OCEAN_BGC"</cppDefs>
        </compile>
        <compile target="static" >
            <cppDefs>"-DUSE_OCEAN_BGC
                -DNI_=1440 -DNJ_=1070 -DNK_=50"</cppDefs>
        </compile>
    </component>
</experiment>
```

- *Inheritance* is a key concept in FRE: an experiment can inherit all of another experiment, and override only those aspects in which it wishes to differentiate itself. This lends itself to an extremely compact description of a series of related experiments, each of which represents a perturbation with respect to some base experiment. One of the FRE utilities, frecanon, can convert the compact form to what is known as the canonical form, in which all inherited values are resolved and independently expressed.
- Within an experiment, information is organized by <component>. Here we see a component called fms, which is the FMS infrastructure itself, and another component called mom4p1, which represents the Modular Ocean Model Version 4 (MOM4) ocean model. For each component, we can have several *methods*, each representing a different stage in the workflow. Typical stages in the workflow include <source>, which prepares the source code; <compile>, which prepares the executable. Other stages in FRE workflow include <input>, which prepares input data; <run> and <postProcess>, not shown in the XML fragment above.
- <source>: each source can come from its own repository. The vc attribute indicates the type of version control, e.g. cvs or Subversion, and root the path to

its associated repository. The <csh> tag can exist for any method, where the user can indicate specialized processing not provided by the standard FRE utilities.

- <compile>: this method provides instructions for compilation. There can be multiple compilation *targets*: each target has different compilation instructions, and thus represents a different realization of an experiment. There can be dependencies among components in compilation, indicated by the requires attribute. The dependency indicates compilation order, for example: the fms component may need to be compiled first.[2]

- <input> (not shown in Code block 1): lists input parameters and datasets for a component. Input parameters are organized into <parameterGroup> s which may be converted by FRE utilities into Fortran namelists, for example. Input datasets follow this syntax:

```
<dataFile source=.. version=.. timestamp=.. checksum=..>
    INPUT/aerosol.nc
</dataFile>
```

As noted here, datasets themselves are under version control. The dataset is delivered to a specific target INPUT/aerosol.nc where it will be read by the model code.

The source could be under hierarchical storage management (see Vol. 4, Chap. 3, Sect. 3 for instance) and involve specialized instructions for retrieval from "deep storage". These instructions are specified in the platform and site configuration sections of FRE, not shown here.

An alternate method of retrieval might be from some remote resource, following the OPeNDAP protocol (Cornillon et al. 1993), for example.

Datasets themselves might be under version control, indicated by the optional version attribute.

Finally, the timestamp and checksum attributes are used to verify data integrity of input datasets.

- <run>: FMS typically runs in single-executable (SPMD[3]) mode. Thus only one component, the top-level coupler component, has an associated <run> method. However, it is easy to express MPMD programming in FRE should one choose to, by assigning <run> methods to multiple components.

Runs are specified in s, each of which constitutes one job on the system. A full model run may be centuries long in model time, however a single run segment is constrained by the limits on the scheduler, how long it permits a single job to remain on the system. Run segments save their states and then resubmit themselves until the end of the simulation. *Production* runs also typically save a

---

[2] FMS is written in F90, where compilation order matters.

[3] SPMD, or single-program multiple-data compiles a multi-component model into a single executable. Other modeling systems, e.g. PRISM, employ the multiple-program, multiple-data (MPMD) mode.

lot of history data and launch a parallel series of post-processing jobs, described below; *development* or *testing* runs save minimal data for immediate analysis. Runs intended for regression testing also list a `<reference>` run, with which they are expected to match answers.

- `<postProcess>`: this method refers to data processing performed upon the output of run or run segment. For example, one might wish to regrid output from the model's native grid onto some standard spatial grid; compute statistics; derive auxiliary variables from saved variables. These computations are data- and I/O-intensive, and do not readily lend themselves to parallelism (see Volume 4 for an in-depth discussion of parallel I/O). Parallelism is at the script level, where processing steps with no mutual dependencies are scheduled concurrently.

  FMS diagnostic output ("history") is organized into subcomponents by time or space sampling. Under `<postProcess>` for the atmos component, one might find subcomponents for monthly and daily data, for instance. Subcomponents may also have associated `<analysis>` methods. Once the post-processing is complete, the data is then processed by visualization and analysis engines to generate graphical output for the user. The ability to run a whole suite of standard analyses on any model run is one of the most compelling features of FRE, the proof of its claim to be an "end-to-end" solution.

This summary look at the structure of a FRE experiment, its components and their methods, provides a glimpse into a functioning and mature *ESM workflow*. Its versatility is shown by the variety of command-line FRE utilities that work off the FRE file:

fremake prepare source and executables for an experiment.

frerun launch and manage a FRE run, in production, development and test modes.

frecheck compare a test run against a reference run, and report discrepancies.

frestatus monitor and report the state of a running FRE experiment.

frepp FRE post-processing and analysis.

freppcheck monitor a post-processing job sequence and report discrepancies and missing data.

frelist list experiments associated with a FRE file, including their inheritance structure.

frecanon convert an experiment expressed with inheritance into a solo FRE experiment with no dependencies.

fredb enter an experiment into the GFDL Curator, a mySQL database of model experiments. This powerful backend system underpins FRE's ability to retrieve prior model configurations.

freversion The FRE syntax evolves as FMS adds new features and components. This utility provides a translator between versions of FRE syntax, and allows users to update their FRE files.

We next discuss lessons learned from FRE, and how workflow concepts might inform the field of Earth system modeling.

2.3 Discussion: Workflows and Curators

Just as FMS provided inspiration and testing grounds for concepts of modeling frameworks that informed the design of ESM frameworks like ESMF and PRISM, so too did FRE lead to efforts to standardize some of these ideas over a larger community, in the Earth System Curator (ESC) project (Dunlap et al. 2008). That project principally followed from FRE's insight that the workflow constitutes the best available description of a model, and thus its *provenance metadata*. The ESC project has had considerable success in specifying formal descriptions of Earth system models constructed from components, and in conjunction with its European counterpart METAFOR (http://metaforclimate.eu), is developing provenance metadata for models participating in large coordinated international experiments, such as the IPCC AR5, scheduled for 2013. By capturing workflow information in a curator database, we will be able to provide some of FRE's services for a wider community.

The architecture of an ESM workflow can also be specified in generic form as in Fig. 2.1. And generic specification systems for scientific workflow such as Kepler (Ludascher et al. 2006) are now becoming popular. Is the time now ripe for an effort to build a standard ESM workflow system? It is tempting to think that a system built on top of a standard framework could indeed implement such a workflow. One could go further and even conceive of model components becoming available as web services that can be directly interfaced from other applications.

While the ESC project does indeed contemplate a Curator Runtime Environment (CRE), there are some caveats to be borne in mind. Principal among these is what is known as the "fuzzy boundary problem" (Sessions 2004–2005). This highlights the fact that the same thing might appear as an *object* (internal data structure) in one application, as a model component in another, and as a web service in a third. In the context of ESMs, we could for instance imagine $g = 9.81$ being a line of code in one model, the value of g being read from a binary file in another model component, and being part of the workflow specification in a third. It is imperative to maintain this degree of flexibility for the developer, and not unduly constrain her in her methods of specification. It is in fact very common indeed for the same application to evolve toward greater formality over time, but that must occur at its own pace.

Nonetheless, the benefits of a more or less formal expression of workflow, and its incorporation into metadata seem to be evident, and we foresee it becoming more prominent in our field with time. Not only does this seem a necessary step in the "operational" use of ESMs alluded to in the beginning of this chapter, there are many new applications that are only dimly seen now, that will become possible with the adoption of formal workflows. For instance, current scientific publications based on model results provide, at best, a reference to the model output data on a server somewhere. We can imagine instead that a scientific article could one day contain a link or URL pointing to a complete workflow file attached to some hardware resource, say a computing "cloud". A scholar reading that article would be able then to manipulate that model directly through the workflow specification, for instance

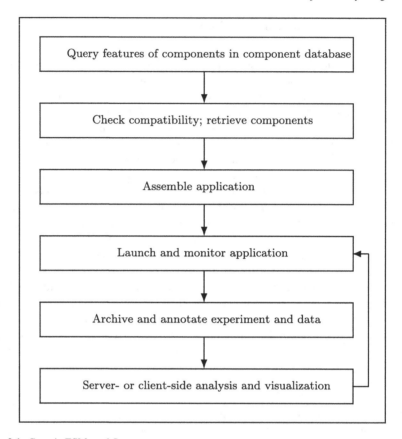

Fig. 2.1 Generic ESM workflow

by rerunning that model varying some parameter! Such an application of workflows would be transformative not only of the field, but of the nature of scientific publication itself.

References

Bray T, Paoli J, Sperberg-McQueen C, Maler E, Yergeau F (2000) Extensible Markup Language (XML) 1.0. W3C Recommendation 6

Cornillon P, G F, J G, G M (1993) Report on the first workshop for the distributed oceanographic data system. Technical report, Graduate School of Oceanography, University of Rhode Island

Dunlap R, Mark L, Rugaber S, Balaji V, Chastang J, Cinquini L, Middleton D, Murphy S (2008) Earth system curator: metadata infrastructure for climate modeling. Earth Sci Inform 1(3–4): 131–149

Larsson C, Wedi N (2006) prepIFS, a software infrastructure tool for climate research in Europe. Geophys Res Abstr 8:04250

Ludascher B, Altintas I, Berkley C, Higgins D, Jaeger E, Jones M, Lee E, Tao J, Zhao Y (2006) Scientific workflow management and the Kepler system. Concurr Comput 18(10):1039

Sessions R (2004–2005) Fuzzy boundaries: objects, components, and web services. ACM Queue 2(9):40–47

Chapter 3
Applying Scientific Workflow to ESM

Ufuk Utku Turuncoglu

As described in the previous chapter, a typical Earth system model (ESM) application manages a series of different tasks, such as configuration, the building and running of the model on various computing resources, and the pre- and post-processing of the input data and model results, and, finally, the visualization of the results. Now-a-days, there are additional tasks concerned with the gathering of metadata about the run environment used, about the model itself and about the input and output data used in a particular run. Due to the complexity of the processes and the multi-component nature of the Earth system models used, each of these tasks requires different levels of expertise and attention. If not supported well, the intricacies of these processes may prevent researchers from focusing on scientific issues, and may make it difficult, or even impossible, to undertake some Earth system science problems.

To help address these issues, a layer of software, or middleware, can be used to simplify the working environment for Earth system modellers. Towards the end of the previous chapter Balaji and Langenhorst mention that a generic workflow approach might be an appropriate way to manage the series of tasks involved. In this approach, the workflow environment can act as an abstraction layer to keep the details of the component tasks, for example, the computing environment, modelling system and third-party applications required, hidden from the scientist-user or modeller. Furthermore, using scientific workflow tools also promotes the use of standardized provenance and metadata collection and archiving mechanisms for ESM applications. This latter aspect is of increasing importance as the volume of data produced from model runs grows in parallel with the development of more sophisticated Earth system models. The metadata information collected plays a critical role when users wish to compare, reproduce, tune, debug, or validate a specific set of simulation runs.

In this chapter, the aim is to give more detail concerning the design, deployment and use of a workflow environment for realistic ESM applications by describing a

U. U. Turuncoglu (✉)
Istanbul Technical University, Istanbul, Turkey
e-mail: u.utku.turuncoglu@be.itu.edu.tr

R. Ford et al., *Earth System Modelling – Volume 5,* SpringerBriefs in Earth
System Sciences, DOI: 10.1007/978-3-642-23932-8_3, © The Author(s) 2012

sample use case that links a well known global circulation model (GCM) with a scientific workflow application.

To set-up and run a multi-component Earth system model like a GCM on a computing resource (for example Grid or cluster system), the user needs to overcome a set of problems including the installation of the model on the specific computing platform (or architecture), the pre-processing and storage of input datasets, the design of a specific ESM case, achieved by the modification of the text-based configuration scripts, the post-processing of model results and the management of the runs. The user also might have to compile and build the model each time the model configuration is changed; as a result, the model runs can include a compilation and build stage. The user may also have to perform repeated steps, such as authentication on remote computer resources, preparation of the required environment for a specific model run, transferring files among storage systems that use different data transfer protocols and authentication systems, and finally running the model. The management of these steps is non-trivial, particularly when considering Grid environments such as Teragrid.[1]

With the work environment that is described in the following sections, each of these individual and repeated tasks can be defined inside the workflow engine as an *actor*, this helps prevent many errors and problems while saving valuable time. In this approach, details of the underlying structures, the computing environment, supporting technologies utilized etc., supporting the scientific application are also hidden from the average Earth system modeller or user.

The rest of this chapter is organized as follows. The next section introduces the realistic use case and its requirements. Then, the components used in the integration of the Earth system model with the scientific workflow application are described. The last section discusses lessons learnt from the experience and provides some thoughts about the future of workflow integration and the implications for the design of the models.

3.1 Use Case and its Requirements

To explain the proposed design approach to create scientific workflow for ESM applications, a realistic use case based on a multi-component and coupled earth system model has been designed. The beta release of the latest version of the Community Climate System Model (CCSM4), which is a subset of the first version of the Community Earth System Model (CESM1), was selected for this purpose. The CCSM[2] is a general circulation model that couples six Earth system model components: atmosphere, land, ocean, sea-ice, glacier via its driver/coupler component. The model includes two different driver components based on the Model Coupling Toolkit (MCT; Larson et al. 2005) and the Earth System Modeling Framework

---

(ESMF; Hill et al. 2004; Collins et al. 2005) which is selected for using in this study. One of the main reasons for using the ESMF enabled version rather than MCT is that ESMF supports the storage of a collection of metadata information about model components (gridded and/or coupler), fields, grid (using GridSpec[3] convention) and about the data exchanged among the components. In this case, ESMF enables one to provide the extra information to develop the self-describing Earth system models (Turuncoglu et al. 2011). The ESMF library also has the capability to export the gathered information in an XML file format which can be processed easily using third party applications (such as scientific workflow applications). In addition to the metadata handling capabilities of ESMF, it also allows one to trigger the ESMF components using simple web service requests and we plan to use this capability to design easy to use, understandable, and much more generic ESM workflows. Detailed descriptions of the ESMF and MCT coupling libraries can be found in Vol. 3 of this series. More information about the CCSM model is given in Vol. 1 of this series.

The design of the use case includes two different Earth system model simulations executing on both a Grid environment (a group of loosely coupled computing resource which are managed by different organizations) and on a standalone cluster system with different horizontal resolutions using the B_2000 model component set. This configuration simulates the present day (20th century) with all model components active (atmosphere, ocean, land, ice and glacier). The configuration options are held fixed except for the horizontal resolution of the atmospheric model component, which is configured as $2.5° \times 0.9°$ and $1.25° \times 0.9°$ for the simulations. The conceptual workflow that demonstrates the whole individual steps to run the CCSM model and analyze the results can be seen in Fig. 3.1. In this context, the details and the duration of the runs are not critically important, since we are mainly interested in exploring how to construct a workflow that includes different model setups, and not in the results coming from the simulations.

It is obvious that the particular requirements of the design of such complex ESM workflow systems depends heavily on the Earth system model(s) involved, on the particular computing environment and on the nature of the scientific problem being studied. However, a number of common requirements that facilitate the interaction among Earth system models and other third party applications or tools (such as scientific workflow applications) can be listed as:

- A generic, extensible and also modular, authentication, job monitoring and submission component.
- Standardized configuration, build and run systems for the Earth system models involved. For example, tools that are similar to "autoconf" enable the creation of standardized configure system for ESMs.
- Self-describing Earth system models that allow one to keep track of the origin and derivation of the specific model run and results generated. The gathered information

---

[3]   http://www.gfdl.noaa.gov/~vb/gridstd/gridstd.html

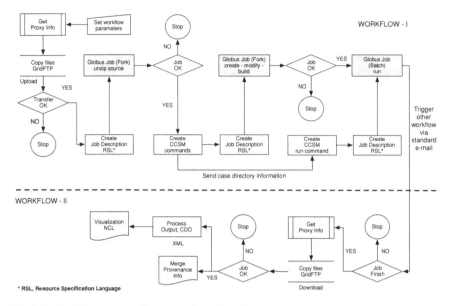

Fig. 3.1 Simplified version of conceptual workflow for use case (markup).

that might be used to compare, reproduce, tune, debug or validate a specific set of simulation runs.

• A computing system that follows basic conventions (such as common environment variables, tools to control the user environment etc.) to create a standardized run environment.

The designed use case addresses the part of the above requirements by modifying the components (build system etc.) of the used Earth system model and integrating it with the *Kepler* scientific workflow application. The following section provides a brief description of the components of the designed use case and of the *Kepler* workflow application and its new module to support ESM applications.

3.2 Components of the Use Case

The components of the typical ESM application contain three basic elements: computing resource, Earth system model and data/metadata portals or centers (Fig. 3.2). The type and architecture of the computing resources that are used to run Earth system models can be very different and particular models are often only portable over a small subset of these. The Earth system model in question can have single or multiple components that a user will have to interact with in order to setup, build and run a specific ESM application. Finally, the data portals can act as the source of the input data or destination of the ESM results. In both cases, it also stores

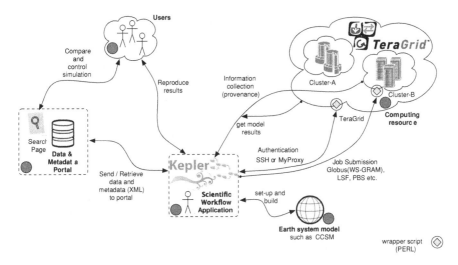

Fig. 3.2 Components and architecture of workflow environment.

metadata or/and provenance information about the data and the simulation itself. It is clear that the design of the modeling environment must take into consideration all of these individual components which will be connected to create the end-to-end workflow (Fig. 3.1).

As can be seen from Fig. 3.2, which depicts a simplified version of the workflow environment, the scientific workflow application resides in the centre of the proposed design and acts as an interpreter to process and transfer the requests among the components. In this design, the users can only interact with the workflow application and data portal.

The rest of this chapter discusses the integration of the scientific workflow application with the computing environment and with the Earth system model used. The design of the scientific data portal and the collection of metadata and provenance information from the Earth system model and from the computing environment are not the subject of this volume. For further reading about data portals in terms of the Grid environment we refer to Vol. 6 in the series.

3.2.1 Computing Environment and Extensions

There are several challenges associated with using *Kepler* to run CCSM on a remote computing resource such as TeraGrid or on a simple cluster that uses a One-Time Password (OTP) for access. In both cases, the user needs to authenticate with the remote system which might reside behind a firewall or Virtual Private Network (VPN), to transfer files or submit the job into queue a system. It is clear that the selected workflow application (*Kepler*) has to deal with all these different kinds of authentication mechanisms and security systems seamlessly. The detailed discussion about using workflow systems in such a kind of secure systems can be found in Podhorszki and Klasky (2008).

The OTP can be basically defined as a password that is valid for only a single session and can be generated by specialized hardware or software. In this way, it adds an extra layer to the overall security of the remote computing resource. The SSH session actor which is a part of the standard *Kepler* distribution, allows one to open a session to a remote computing resource. The SSH session remains open until the execution of workflow finishes and the user can execute different commands without entering the password. In this case, the user has to pass the generated password manually at the beginning of the workflow execution.

After authentication on the specific resource, it is necessary to create a suitable environment for the simulation, which is unique and may require third-party applications to run (such as compilers, libraries, visualization software, etc.). The workflow application might also need a wide range of information about the target computing environment, such as the location of a parallel file system or of other libraries and applications that are required.

Normally, all this information must be defined in the user's environment on the remote resource. This implies that the user must know the location, or path, of the required applications or, at a minimum, their names. It is this process that can be automated in order to create a standardized environment for the workflow. To deal with these issues, in *Kepler*, a wrapper Perl script is created to define the necessary application environment variables (e.g. PATH) and to replace specific workflow variables with site-specific definitions before the job executes (e.g. TG_CLUSTER_HOME, TG_CLUSTER_PFS, ESMF_LIB_DIR). The wrapper script supports common tools that are used to manage the environment variables, such as, for example, Argonne National Laboratory's SoftEnv applications[4] and Modules.[5] The operating logic of this script is simple: in the beginning, the script tries to find the location of the SoftEnv or Modules application on the remote machine and then loads this location into the users environment. Next, it searches for the required applications or libraries in the SoftEnv or Modules database. For those that are found, the script loads its location into the users environment. The output of the script is an array of commands which can be added to the beginning of the command that will be used to execute the model run.

3.2.2 *Kepler Scientific Workflow System and Extensions*

A scientific workflow system creates standardized interfaces to a variety of technologies and automates the execution and monitoring of a heterogeneous workflow (Altintas et al. 2006). A workflow system is also defined as a problem-solving

---

[4]   http://www.mcs.anl.gov/hs/software/systems/softenv/softenv-intro.html

[5]   http://modules.sourceforge.net/

environment that simplifies complex tasks by creating meaningful, easily under-standable sub-tasks/modules and the means of combining them to form executable data management and analysis pipelines (Bowers and Ascher 2005; Ludäscher et al. 2005). There are several scientific workflow applications in existence such as *Kepler* (Ludäscher et al. 2005), *Taverna* (Oinn et al. 2004), *Triana* (Majithia et al. 2004) and *VisTrails* (Bavoil et al. 2005). In contrast to standalone scientific workflow systems, the Linked Environments for Atmospheric Discovery (LEAD; Plale et al. 2006) project demonstrates how workflows can be used to solve problems specific to earth system science by integrating various technologies such as web and Grid services, metadata repositories, and workflow systems. The LEAD scientific gateway merges a sophisticated set of tools to enable users to access, analyse, run, and visualize meteorological data and forecast models, facilitating an interactive study of weather. Another example, similar to the LEAD project, aims to integrate the Grid ENabled Integrated Earth system modelling (GENIE) system with a workflow application in order to manage the series of operations in ESM simulations. This system aims to define MATLAB-based scripts with their workflow representations by using the Windows Workflow Foundation (WF), which is a technology supporting the rapid construction of workflow-enabled applications. For further reading about the GENIE project we refer to Vol. 1 of this series and to Chap. 7 of this volume.

The *Kepler* scientific workflow environment is selected for this study because it is free, open source and platform independent software that has a modular design and support for different models of computation, such as the execution of individual tasks in a threaded (parallel) or sequential way. It also allows users to design scientific workflows and execute them in local and remote computing environments using either a graphical user interface or via the command line. *Kepler* is mainly based on the *Ptolemy II* (Eker et al. 2003) project from which it inherits modelling and design capabilities.

The *actor*-oriented design of *Kepler* enables users to create hierarchically struc-tured or nested scientific workflows. Actors are responsible for performing actions on input data to produce output. They can have multiple input and output ports that pass streams of "data tokens" to other actors. The overall model execution is controlled by a separate component (a kind of specialized actor) that is called a *director*. Both actors and directors have parameters that define specific component behaviours. *Kepler* components (actors and directors) and the workflows created can be defined, stored or exchanged in a standardized Extensible Markup Language (XML) format, called the modelling markup language (MoML; Lee and Neuen-dorffer 2000). MoML enables users to define workflow elements such as actors, directors, connections, relations, ports, and the parameters required to build models and workflows.

The main window and components of the *Kepler* application can be seen in Fig. 3.3. The sample workflow, which is shown in the figure builds and runs a simple Message Passing Interface (MPI) application on a Grid environment using a bunch of specialized *Kepler* actors. *Kepler* is a Java-based scientific workflow application and new actors can be created easily by using low-level Java classes. Each *Kepler* actor contains, at the minimum, three different components: input ports, output ports

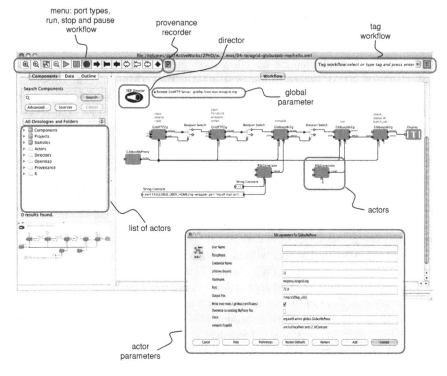

Fig. 3.3 Main *Kepler* window with major sections annotated.

and parameters. To adapt an Earth system model code to run in *Kepler*, one must first analyse the steps that are necessary to compile, configure and run the model and the relations between them. Later, these steps are converted into their actor representations.

For this use case study, a new domain specific *Kepler* module, which is called "earth" is developed. The latest version of this module can be found under the ESMF Contributions on the SourceForge[6] website (Turuncoglu et al. 2011). The new module includes many ESM-specific actors that are mainly responsible for modifying the configuration of the model components (such as WRF,[7] ROMS[8] and CCSM) and building the source code. The module also includes the computing environment and visualization-specific actors, demonstration workflows, and wrapper and provenance collection scripts. The newly designed actors can be grouped as follows:

- *Computing environment-specific actors*: a new set of actors is created to support authentication, file transfer, job submission and monitoring in Grid environments.

---

[6] http://esmfcontrib.cvs.sourceforge.net/viewvc/esmfcontrib/workflow/

[7] http://www.wrf-model.org/index.php

[8] http://www.myroms.org/

These newly designed actors have a crucial role in the overall workflow environment because, at the time of this study, *Kepler* Grid computing actors do not support the following functions: authentication via MyProxy[9] server, submitting Web Service interface enabled Grid Resource Allocation and Management (WS-GRAM) jobs and automatic creation, of XML-based RSL (Resource Specification Language) job definition scripts. RSL is defined as a language that is used to specify resources needed by a Globus job.The new actors implement all of these.

- *Earth system model-specific actors*: to enable CCSM based workflows, custom *Kepler* actors were designed to handle the configuration, build and run stages of the CCSM model. Fortunately, the new version of CCSM (CCSM4) uses XML as a standard data format to define the model configuration files and parameters rather than raw American Standard Code for Information Interface (ASCII) files. The new configuration structure enables us to integrate all stages easily with *Kepler* actors, using the Java XML Application Programmer Interface (API). In this manner, actor parameters are generated dynamically to capture revisions or modifications of the CCSM configuration files without modifications of previously developed *Kepler* actors.

- *Post-processing-specific actors*: to process model output from the two applications, the Max Planck Institute's Climate Data Operator (CDO) and the National Center for Atmospheric Research (NCAR) Command Language (NCL), which are widely used by the Earth system science community, were incorporated into the workflow environment. In the CDO actor, an Web Ontology Language (OWL) knowledge base is used to store the relationships of the different processing options and their parameters. The OWL is defined as a knowledge representation language to create new ontologies. The actor simply queries this knowledge base using the SPARQL protocol and RDF Query Language (SPARQL) language to create the parameters needed for specific processing commands.

In addition to newly designed actors, the *Kepler* application itself is also modified to support the triggering of the follow-up workflows automatically. Getting results from Earth system models, especially climate models, can take days and even months. It does not make sense to handle this kind of simulation in a start-run, execute-run, examine-run fashion because of those time scales. Furthermore, creating a giant workflow that includes the whole range of modelling tasks is not a scalable and efficient solution when evaluated within the modular design paradigm. For these reasons, users tend to separate the whole workflow into small and meaningful distinct parts. The conceptual workflow that was shown in the previous section (Sect. 3.1) can be seen as an example of this approach. The first part of the workflow configures and runs the simulation, collects metadata, and then merges and sends the collected metadata into an archive (see Figs. 3.1, 3.2). The second part is responsible for the post-processing of results. In the proposed approach, the upstream workflow, or the Earth system model itself, can control the execution order of distinct workflows by sending simple e-mail messages that contains the job description information in

---

[9] http://grid.ncsa.illinois.edu/myproxy/

```
<?xml version="1.0" encoding="UTF-8"?>
<workflow>
  <!-- workflow and its location-->
  <name>07-get-output.xml</name>
  <directory>/Volumes/dali/system-backup/workspace/kepler.modules/earth/demos</↵
        directory>
  <!-- workflow parameters -->
  <parameters>
    <gridftp_srv>gridftp.frost.ncar.teragrid.org</gridftp_srv>
    <output_dir>/Users/turuncu/Desktop/output-wf-beta16</output_dir>
    <remote_case_dir>/ptmp/turuncu/kepler.case.BTR1.f19_g16</remote_case_dir>
  </parameters>
</workflow>
```

Fig. 3.4 Example job description XML for follow-up workflows

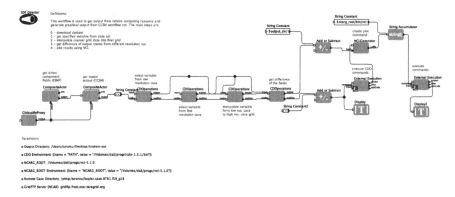

Fig. 3.5 Screenshot of the *Kepler* follow-up workflow that is triggered by XML (see Fig. 3.4).

an XML form (Fig. 3.4). The pre-defined e-mail account used is simply checked at certain intervals and parsed by the modified *Kepler* loader to trigger the appropriate follow-up workflows automatically when the upstream task finishes. It is clear that the vulnerability and security of the computing resource or the server that workflow application resides in becomes an important issue in this method because of the potential for malicious e-mail attacks. To prevent a possible security hole, the messages can be encrypted using gnuPG[10] or similar security methods.

As can be seen from Fig. 3.4, the job definition file contains information about the follow-up workflow by using "name" and "directory" XML tags and its parameters which resides inside the "parameter" XML tag. In this way, the behavior of the workflow might be controlled from outside the workflow environment or the upstream task. The sample XML file is used to trigger a file transfer workflow (named as 07-get-output.xml) to copy CCSM results from TeraGrid to a local machine via the GridFTP protocol (Fig. 3.5).

In this use case study, *Kepler* is used to merge the different components of the conceptual workflow and to construct a prototype end-to-end modelling system. The

---

[10]   http://www.gnupg.org/

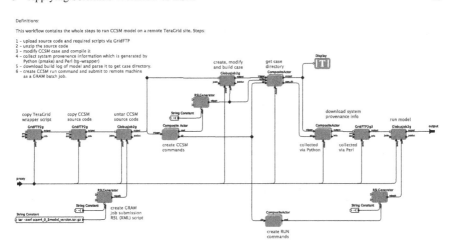

Fig. 3.6 Screenshot of the *Kepler* version of CCSM workflow (content of the main composite actor, TeraGrid).

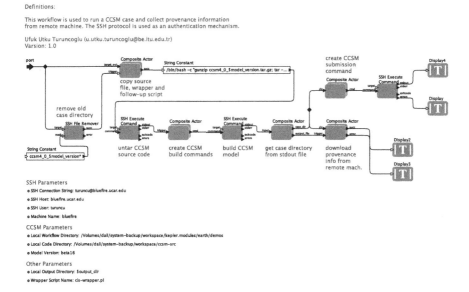

Fig. 3.7 Screenshot of the *Kepler* version of CCSM workflow (content of the main composite actor, NCAR's Bluefire cluster).

workflow creates, modifies, builds, and runs the fully active CCSM model on two different computing resources: the TeraGrid (Fig. 3.6) and a cluster that uses the SSH protocol and a one-time password authentication mechanism (Fig. 3.7).

The top level CCSM workflow contains two composite actors (which contain sub-workflows) that are controlled by a *Kepler* PN (Process Networks) director because of the independent nature of the task. The content of one of the composite actors is shown in Fig. 3.6. Each composite actor (containing more than 100 individual actors) is responsible for creating, modifying, installing, and running CCSM use cases that are configured to run at different spatial resolutions. The composite actors contain an inner Synchronous Data Flow (SDF) director because the configure, build and run stages must be executed in sequential order. In *Kepler* terminology, this kind of composite actor is called an *opaque* composite actor, because its internal structure is neither visible nor relevant to the outside director.

To run the same workflow on a cluster system, the user only needs to change the Grid environment-specific actors to their cluster environment equivalents. For example, the authentication specific "MyProxy" actor is replaced by an "SSH Session" actor to keep open a session until execution of the workflow finishes (Figs. 3.6, 3.7). As can be seen from the example, the modular nature of the *Kepler* scientific workflow environment simplifies the process of new workflow creation and facilitates the flexible interoperability among its components.

3.3 Conclusion

The use case described shows that the modelling system developed facilitates the integration of the different components of the proposed workflow environment and also enables an easy to use and efficient working environment. The prototype workflow environment presented is capable of running on different types of high performance computing resources through the use of CCSM's configure, build and run scripts and modular authentication, job monitoring and submission *Kepler* actors. Nevertheless, the designed CCSM specific actors are still tightly coupled to the ESM and its build system.

One of the basic problems encountered when integrating an Earth system model with a scientific workflow application is that Earth system models mostly use non-standardized, text-based (i.e. ASCII) model configuration/parameter definition systems, and these configuration files are parsed by model code to create different modelling cases. It is clear that this may be a source of interoperability problems. The CCSM model that is used in this use case study takes advantage of XML standards to define configuration parameters, and their options, to standardize the model configuration interface. As a result of using XML standards to define model configuration instead of text-based systems, the overall integration process of the workflow application is simplified, and adaptability to different versions of the model being used is achieved. Despite all of the apparent advantages of the XML standard, it is not yet gaining wide acceptance with ESM developers because of the development overhead to implement such types of configuration system. A *wrapper* layer might be designed to move back and forth between non-standard ASCII and structured XML-based configuration files to address this issue for Earth system models that use old fashioned configuration files. The design and implementation of such a kind of *wrapper* layer can be difficult when we consider the wide range of complex parameter configuration files that are actually used by ESMs. While it is difficult to integrate the Earth system models that have non-standard interfaces, such as WRF and ROMS, with scientific workflow applications, the proposed work environment has ESM model specific scripts to handle these technical issues.

The problem of collecting metadata and provenance information about ESM simulations that are triggered by the workflow system is out of scope for this volume, but the prototype version of work environment presented also addresses this problem by using a new concept: the self-describing Earth system model. As mentioned in Sect. 3.1, self-describing Earth system models play a crucial role when the user wants to compare, reproduce, tune, debug, or validate a specific set of simulation runs. In the use case presented in this chapter, the combination of the CCSM model and a set of provenance collection scripts enable the export of information about the Earth system model and the target computing environment. For this purpose, ESMF Attributes are used to produce the component-level CCSM model metadata. The experiments show that the proposed architecture is suitable for collecting detailed and standardized provenance information to create self-describing Earth system models by using *Kepler*'s workflow provenance recorder module and the newly created script tools.

To address the problem of creating a standard interface to computing resources and collecting system level metadata, the new set of scripts have been created based on a modified version of the Oak Ridge National Laboratory (ORNL) and North Carolina State University (NCSU) "pymake" tool. "pymake" is an independent or stand-alone component which gathers information about the build process and it is loosely coupled into the proposed work environment. Using web services, for example, third-party applications such as scientific workflow systems can query the user and system-level information from a remote computing resource and use the information to define the specific set of tasks required before submitting the job onto the remote machine. In this manner, the model build and run stages might be simplified.

3.4 Future Directions

In the previous sections, research efforts in the application of scientific workflow to ESM have been described. In this section, some possible future research directions are proposed.

The main disadvantages of the proposed workflow implementation is that the ESM specific Kepler actors (CCSM build, run and configure actors) are tightly coupled to the particular model and a user would need, to design a new set of actors in order to port the designed work environment or to apply a similar methodology to other ESM models. It is obvious that the ability to seamlessly represent an ESM outside of its workflow environment can help to improve the portability of the overall system. Some research efforts on integrating Earth system models with third party scientific workflow applications have focused on utilizing the web services approach to trigger and set up Earth system model components, rather than creating Earth system-specific *Kepler* actors to overcome this issue. It is clear that the web services approach might be a more generic and scalable solution, not withstanding the potential impact on performance. With web services, the interfaces of the Earth system models might be able to be standardized. Further, models that have the ability to process SOAP (Simple Object Access Protocol) requests would support the development of generic scientific workflow systems. To this end, the new, prototype version of an ESMF socket interface that allows any networked model or coupler component to be available as a web service, might be used. The OpenMI[11] software component interface provides an alternative to the ESMF web service component. In both case, the components of the model are designed to be available as a web service.

Another research goal would be to design self-describing and standardized Earth system models that follow common conventions and technologies. With this approach, tightly coupled end-to-end modelling systems can be designed and interoperability of the model with third-party applications, such as data or metadata portals, science gateways and other workflow environments, may be achieved. In this context, the scientific workflow applications might be used to automate the processes of collecting metadata and provenance information from the Earth system model, computing environment and data files etc., enabling more flexible and rapid Earth system science experiments. The current version of the workflow environment is able to collect a small set of provenance information (system and data provenance) using specialized scripts. However using generic tools that are designed to collect different types of provenance information might enable the collection of much more detailed information from the model, the running environment and the workflow system itself. The Provenance-Aware Storage System (PASS, Muniswamy-Reddy et al. 2006), Earth System Science Server (ES3, Frew et al. 2008) and *Karma* provenance collection framework (Cao et al. 2009) can be used for this purpose.

References

Altintas I, Barney O, Jaeger-Frank E (2006) Provenance collection support in the kepler scientific workflow system, Provenance and Annotation of Data, pp 118–132. doi:10.1007/11890850_14. http://dx.doi.org/10.1007/11890850_14

Bavoil L, Callahan SP, Scheidegger CE, Vo HT, Crossno P, Silva CT, Freire J (2005) Vistrails: enabling interactive multiple-view visualizations. In: IEEE Visualization, IEEE Computer Society 18

Bowers S, Ascher B (2005) Actor-oriented design of scientific workflows. In: In 24st International conference on conceptual modeling, Springer, pp 369–384

Cao B, Plale B, Subramanian G, Robertson Ed, Simmhan Y (2009) Provenance Information Model of Karma Version 3. In: Proceedings of the 2009 Congress on Services-I. IEEE Computer Society, Washington, pp 348–351. doi: 10.1109/SERVICES-I.2009.54. http://portal.acm.org/citation.cfm?id=1590963.1591574

Collins N, Theurich G, Deluca C, Suarez M, Trayanov A, Balaji V, Li P, Yang W, Hill C, Da Silva A (2005) Design and implementation of components in the earth system modeling framework. Int J High Perform Comput Appl 19(3):341–350. doi: 10.1177/1094342005056120

---

11   http://www.openmi.org/reloaded/

Eker J, Janneck J, Lee EA, Liu J, Liu X, Ludvig J, Sachs S, Xiong Y (2003) Taming heterogeneity— the ptolemy approach. Proceedings of the IEEE 91(1):127–144

Frew J, Metzger D, Slaughter P (2008) Automatic capture and reconstruction of computational provenance. Concurr Comput: Pract Exper 20:485–496. doi:10.1002/cpe.1247. http://dx.doi.org/10.1002/cpe.1247

Hill C, DeLuca C, Balaji V, Suarez M, da Silva A (2004) The architecture of the earth system modeling framework. Comput Sci Eng 6(1):18–28. doi:http://doi.ieeecomputersociety.org/10.1109/MCISE.2004.1255817

Larson J, Jacob R, Ong E (2005) The model coupling toolkit: a new fortran90 toolkit for building multiphysics parallel coupled models. Int J High Perform Comput Appl 19(3):277–292. doi:http://10.1177/1094342005056115

Lee EA, Neuendorffer S (2000) Moml-a modeling markup language in xml-version 0.4. Tech. Rep. UCB/ERL M00/12, EECS Department, University of California, Berkeley. http://www.eecs.berkeley.edu/Pubs/TechRpts/2000/3818.html

Ludäscher B, Altintas I, Berkley C, Higgins D, Jaeger E, Jones M, Lee EA, Tao J, Zhao Y (2005) Scientific workflow management and the kepler system. Concurr Comput Pract Exper 18(10):1039–1065

Majithia S, Shields MS, Taylor IJ, Wang I (2004) Triana: a graphical web service composition and execution toolkit. In: ICWS, IEEE Computer Society, pp 514

Muniswamy-Reddy K, Holland DA, Braun U, Seltzer M (2006) Provenance-aware storage systems. In: Proceedings of the annual conference on USENIX '06 Annual Technical Conference. USENIX Association, Berkelay, p 4. http://portal.acm.org/citation.cfm?id=1267359.1267363

Oinn T, Addis M, Ferris J, Marvin D, Senger M, Greenwood M, Carver T, Glover K, Pocock MR, Wipat A, Li P (2004) Taverna: a tool for the composition and enactment of bioinformatics workflows. Bioinformatics 20(17):3045–3054 URL http://bioinformatics.oxfordjournals.org/cgi/content/abstract/20/17/3045

Plale B, Gannon D, Brotzge J, Droegemeier K, Kurose JF, McLaughlin D, Wilhelmson R, Graves SJ, Ramamurthy M, Clark RD, Yalda S, Reed DA, Joseph E, Chandrasekar V (2006) Casa and lead: adaptive cyberinfrastructure for real-time multiscale weather forecasting. IEEE Computer 39(11):56–64

Podhorszki N, Klasky S (2008) Workflows in a secure environment. Distributed and Parallel Systems pp 143–153. doi:10.1007/978-0-387-79448-8_13

Turuncoglu UU, Murphy S, DeLuca C, Dalfes N (2011) A scientific workflow environment for earth system related studies. Computers & Geosciences 37(7):943–952. doi:10.1016/j.cageo.2010.11.013. http://www.sciencedirect.com/science/article/pii/S0098300410003808

Chapter 4
Configuration Management and Version Control in Earth System Modelling

Mick Carter and Dave Matthews

4.1 Introduction

This chapter discusses the process of developing and constructing Earth system models (ESMs). The construction of an ESM is fundamentally an action of *configuration management* (CM), the process of handling change in software systems which may be large and complex. Essentially, configuration management involves identifying, at a level suitable for management, the components of a software system, controlling changes to components over time (i.e. *version control*), providing traceability of changes made and supporting the maintenance of auditable versions of the software system. Configuration management typically uses the language of trunks, branches, commits, merging and releases and we assume the reader has some familiarity with these terms. For readers unfamiliar with this terminology, a good introduction may be found in CM (2010),[1] Pilato et al. (2008).

In Earth system modelling, the aim of configuration management and version control is to produce an integration of an ESM that:

- Conforms to the experimental design.
- Can be reproduced.
- Has a well defined pedigree (what it is based upon).
- Can be recognised and understood in the context of other integrations.

Further, configuration management supports the ability to trace changes in the code for reasons of comparison and debugging.

---

[1] http://en.wikipedia.org/wiki/Configuration_management

M. Carter (✉)
Met Office Hadley Centre, Exeter, UK
e-mail: mick.carter@metoffice.gov.uk

D. Matthews
Met Office, Exeter, UK
e-mail: david.matthews@metoffice.gov.uk

R. Ford et al., *Earth System Modelling – Volume 5,* SpringerBriefs in Earth System Sciences, DOI: 10.1007/978-3-642-23932-8_4, © The Author(s) 2012

To explain the process, we will look at the task of configuration management from the point of view of a number of *roles* as defined at one institution that carries out ESM: the Met Office in the U.K. The various roles will be considered and the challenges that they face will be discussed, but it should be noted that they overlap. The order in which we describe the roles is somewhat hierarchical in that some roles predominantly depend on others, although there is always a degree of two-way interaction. We discuss the processes to support configuration management currently in use at the Met Office and give a critical review of their known limitations, offering suggestions for future improvements. As in all organisations, these processes need to continue to evolve to meet new challenges. A description of the configuration management tools discussed here can be found in Matthews et al. (2008).

It is useful to remember that the disciplines of configuration management are needed to allow the reproducibility of results which is at the heart of the scientific method. In the Met Office, reproducibility of results is taken to the extreme of requiring 'bit level' reproduction of the output from integrations, ideally even across different processor counts. No configuration management system can help with maintaining such bit reproducibility across platform changes or compiler changes and this has to be managed by comparison methods. At the Met Office, a key method used is the analysis of divergence of the early part of the solution from perturbed initial states with the divergence of the solution on the new platform and this is followed by longer integrations and a comparison of the climate characteristics with a baseline run. This latter exercise is particularly challenging because of natural climate variability. Thus, ultimately, when targeting execution on a particular computer, the configured software system includes not only source code but also compilers, build scripts and any other software tools employed in the building and running of an ESM.

The roles that will be considered are:

- The software infrastructure developer.
- The scientific developer.
- The model configuration developer.
- The configuration manager.
- The model release manager.

Individuals at the Met Office may have more than one role but each role has its own expectations and responsibilities. These are explored in the following section.

As this chapter of the book uses the Unified Model (UM) as an example, it is worth providing a brief description of that system, based on the repository and the repository's project structure as well as a description of what the Unified Model User Interface (UMUI) provides.

The whole UM subversion repository shares the same Trac system with one set of tickets and milestones for all projects within the repository. The projects in the UM repository have aligned release dates, when validated branches are merged onto the trunk. There are separate projects within the UM repository for (i) the UM code and scripts, (ii) the UMUI, (iii) documentation and (iv) various supporting libraries. The UM project is further sub-divided for reasons of convenience and has directories for scripts, utilities, atmospheric code and build configurations.

The NEMO ocean model, used in our coupled configurations, has its own repository and soon, the Joint UK Land Environment Simulator (JULES) land surface scheme will also be in its own repository.

This set of repositories covers all UM configurations from coupled climate earth system models to atmospheric mesoscale forecast models. The model configurations are defined by input values, logical switches, option choices as well as scripts and environment variables.

In climate mode, the UMUI allows the user to set up a whole climate run, which runs in segments that each submit the next segment of the run. The UMUI allows the user to define each scientific parameter, diagnostic requirements, technical control for components and the whole run (e.g. dumping frequency, length of each run segments and domain decomposition). The UMUI also collects information about the target platform and the user's account on the platform. Through the UMUI, the user can define the compilation and code used. The user can select a number of code branches to use as long as those branches do not clash, in which case a merged branch will need to be created. Each of these branches contains a logical change, or set of changes, to a given version of the UM. The branches may be candidates for a future UM release or something more experimental.

There is a UMUI release for each released version of the UM and each version of the UM can be set up to define any possible configuration of the UM.

4.2 Role 1. The Infrastructure Developer and Supporter

The UM is the numerical modelling system developed and used at the Met Office for all of its weather forecasts Numerical Weather Prediction (NWP) and climate predictions. There is a technical team at the Met Office that develops and supports the UM system. There are various reasons for the development of infrastructure as follows:

- To provide a supported environment, with a well defined code base and associated software.
- To support changes in the environment, or alternative environments (environments include: computer, compilers, operating systems, archive systems, post processing and monitoring systems etc.).
- To make the ESM more efficient.
- To make the ESM more easy to use.
- To support the technical needs of new or changing scientific components (e.g. coupling systems, integration of new models).
- To provide the scientists with more information (e.g. new diagnostic systems).

From a version control/configuration management point of view, people in this role work in much the same way as many software development projects. Software is developed and targeted at a particular release. However, they need to address the challenging requirements of their user community, the scientists running ESMs, where there is a tension between the need for stability[2] and the need to react to changes in the environment outside the end user's control. Such changes may include support for new computers, new code optimisations, new diagnostics or new usability features.

The Met Office solution has been to provide a monolithic release of consistent software and users move to new versions when they are ready. Further, a release of the UM will preserve scientific formulations, as well as adding new variations because the release needs to be used by a variety of user groups that change the science base on different timescales. The primary aim has been to allow configurations to generate exactly the same results across UM releases for key configurations. This scientific stability is required to encourage users to move to new releases that could provide more functionality, integration with new technical infrastructure or better performance. Although scientifically safe, there are several known weaknesses in the current system:

- Technical changes, for example to cope with a new archiving system, that we would want all users to use, are not picked up transparently by the user. The user would need to apply a branch to make the change. This is because technical code is in the same code base as scientific code and therefore subjected to the same principle of code stability that is required for scientific code. So, system changes that we may want to be mandatory have to be applied electively by users in the same way as scientific changes need to be applied electively by users to ensure reproducibility of the science.
- The effort involved in moving to a new release is significant because the science has to be verified. Major configurations are verified centrally, but most users run derivations which require individual attention.
- There is increased complexity because the UM has to support multiple science configurations.
- Users may not be able to wait for a new release.

The current trend is to evolve the system to be less monolithic. One aim would be to provide technical infrastructure separately from a UM release. The UM does this to an extent today, with the FCM system[3] for source code control and compilation and the Ocean Atmosphere Sea-Ice Soil (OASIS) coupler (Valcke 2006) being on a separate release cycle. We are used to coping with interfacing with external systems that need to change (libraries, compilers, post-processing, archive servers). However, there is much more that could be done (e.g. for diagnostic systems and archiving clients). Further, there would be much more to gain if the interfaces to these systems were stable. A less monolithic system would allow users to more easily and quickly take on technical changes without disturbing the scientific results. It has been the

---

[2] Changing versions can be a slow and painful process because of the need to verify that the science has not changed.

[3] FCM: Flexible Configuration Management system (Matthews et al. 2008). FCM was developed at the Met Office. It is built on top of Subversion (Pilato et al. 2008) and is designed to encourage a particular development model that is suitable to this type of scientific development.

aim of the FLUME project (see Volume 1 in this series) to increase modularisation and provide cleaner interface layers between the components.

The infrastructure developer needs configuration management tools to support their development activity and, at the Met Office this activity is supported by the FCM system.

4.3 Role 2. The Scientific Developer

The scientific developer has the job of developing new or improved scientific code and integrating it into the model. Their starting point will normally be a defined model configuration provided by a configuration manager (see Role 3 in Sect. 4.4).

The job of integrating the scientific change can be easy or hard depending on the amount of interaction required on the following scale:

- No interaction with the infrastructure. The scientific code is self-contained. This is seldom the case, as the scientific code often needs some sort of user control or has some new diagnostics that require integration into the UM User Interface (the UMUI).
- Standard supported interaction with infrastructure. For example, this could include the addition of new diagnostics.
  Diagnostic are presented as options to the user through the UMUI. The UMUI is aware of the diagnostics through metadata records. Today, the user cannot change those records within a configuration management system. Because the UMUI cannot be changed by the user of a given UM release, in contrast to the code which can be modified, it is hard to provide a user interface that is consistent with the set of science code changes that the user is developing.[4] In the case of new diagnostics, the system allows the user to provide their own files of metadata records that define the new diagnosis that they have coded. The UMUI is able to read these metadata files and present the user with the appropriate control to allow the diagnostics to be used. However, these metadata files defining new diagnostics are not automatically under change control. Because the diagnostics are uniquely defined by an enumeration system, references in the metadata in the archive need to be checked against records of the job definition (this failing is made more problematic because the file formats in use currently define diagnostic types by a code number rather than by human readable name, and this code number is only defined by the user-provided file which is not under change control). The system does provide a mechanism for recording this information.
  This example illustrates that the proper configuration management of metadata within the system, defining inputs and diagnostics available to be set within the user interface, is key to proper governance of the output diagnostics. Interaction

---

[4] The UMUI is fixed at each UM release because, to date, no way has been found to implement a facility for making changes to the interface in real time for an interface that is a fat client running in a distributed environment.

between the user interface and scientific changes can also cause other complications. For example, if the scientific developer wants the user to be able to make some selection (e.g. of an option within an algorithm or the value of a tuneable parameter) then, again, they would really want the user interface to reflect the code branch in development without having to wait for a new UMUI reflecting the code change at the next UM release. One way to get around this is to set a default value, but different users require different defaults (e.g. a tuneable parameter may be resolution dependent or an option is set one way for forecast models and another way for climate models). This is not easy to achieve and indirect methods are used to record the configurations.

The basic problem here is that scientific changes may be required to be used by a range of end users before they have been formalised in a UM release. An example of this is where those developing new scientific configurations need to test new science as it is being developed, and development activities may be going on, in parallel, across a number of different configurations. This is because the UM is used for both climate and weather forecasting, at a number of resolutions and with a number of combinations of model components.

- Changes that require extension to the infrastructure. Here, Role 2 is reliant on work done under Role 1 to extend the infrastructure. Most changes require this to some degree as the user interface will often need to be extended to allow new code to be controlled, including the requesting of (new) output made available by the new code.

The scientific developer also has to consider users of the UM who want scientific stability. This is a non trivial matter since the amount of code change is large—the Met Office's UM has changes to about 30% of the lines of code each year. In the Unified Model, the provision of stability has been achieved in various ways:

- There are multiple releases of the UM in existence and these are used in many different climate and NWP configurations. New science is introduced formally on the trunk of a new UM release (typically, the next release following completion of development) but can be accessed via branches against the (older) base release from which it was developed. Some science can be dropped at a new release but this is avoided when the science is actively used. Keeping older science in new releases reduces the number of UM versions that need to be supported.
- New science can be introduced without precluding the use of old science via run-time options that selectively activate the relevant blocks of code. However, this method cannot always be used as the changes can be too complex to handle with this method.
- As an alternative, the UM supports the ability to have multiple copies of a scientific section (e.g. convection is a section of the atmospheric model) within a single version of the UM. This facility supports quite different formulations of the science through use of alternative routines but adds complexity.

The scientific developer needs configuration management tools to support their development activity, in the same way as was seen for the infrastructure developer.

However, the work of the scientific developer is more likely to span multiple releases of the code base, and the need for tools to support an extended development cycle (primarily tools to support merging of development branches) is more important.

4.4 Role 3. The Configuration Developer Team

The configuration developer has the job of combining scientific components, defining options and parameter values and other input data to meet a specific scientific aim. Typically, this person (or, more typically team) will develop a base configuration from which a number of configurations will then be developed. The task will involve a lot of experimentation before the configuration is finalised. Taking a lead on best practice, and acting as a guardian for configuration management within the team is the configuration manager. It has been found important that key climate configurations at the Met Office each have a configuration manager.

The configuration developer will follow a scientifically-motivated process, rather than a technically motivated process. They will need to keep records of what they have tried and relate them to results obtained. Configuration management processes can help here and the Met Office has recognised a need to further develop the practices and tools for the configuration developer to reduce the effort required managing the complexity. For example:

- The configuration developer will need to combine a number of branches[5] of development relating to different scientific developments to check their validity. These codes reside on branches rather than the trunk because they have not been proven. The configuration developer needs to ensure that they take well defined revisions of such branches. The FCM system developed by the Met Office provides support for this process via its extract system. This system interfaces with Subversion and is able to combine code together from multiple branches to be built for a particular experiment (so long as the code changes do not clash). Without this system, developers would be more likely to combine together changes in working copies and then build from this code which is not an easily repeatable process, and which is, therefore, undesirable.
- Scientific code may need to be changed as a result of the experimentation done in the configuration development process. This may be done by the configuration developer themselves or, on referral, by the scientific developer. Change control tools help here too, but this process relies very much on human coordination to ensure that the right version of the code eventually is applied to the trunk for the next release and careful release testing against developing configurations is required.
- The team of configuration developers will also make changes, under expert advice, to settings and adjustable parameters to tune their models. At the Met Office this is often done through the user interface, which currently has no concept of

---

[5] Branches are amended copies of the entire project and represent a single logical change (such as an amended version of the radiation algorithm) or a collection of logical changes.

revision control for these settings. Configuration developers replicate the functionality through use of multiple job definitions and associated project documentation. The Met Office aims to develop the user interface both to allow model definitions to be held under revision control and to support its extension to cope with the user's need to add new input parameters, not yet supported by the UMUI.

- A UM job defined through the user interface points to resources other than code, such as ancillary files (which contain boundary conditions and forcing data). Configuration managers manually create repositories of ancillary files and the Met Office is developing a mechanism for registering standard ancillary files into a repository with sufficient metadata and base data to allow a developer to reproduce them, or more usefully, produce variants of them. This will also benefit remote users of jobs as standard ancillary files will be more easy to exchange.
- At the Met Office, a best-practice standard has been developed for validating models and this is captured in a set of Validation Notes. Use of this practice is a help and provides a level of consistency. Further benefits can be gained by integrating such a system with configuration management of UMUI jobs and descriptive metadata for jobs.

Further, during the development of new configurations, some of the code base being used may be in an experimental state as a result of other, on-going, developments. So the configuration developer needs tools to manage collections of codes that are being developed by other people in the configuration developer role. It is important that these codes are well defined (i.e. that they are formal release points of a branch). The implementation of FCM at the Met Office supports, in an integrated way the naming conventions and other constructs to support this activity. In particular, two useful constructs are:

- *Package branches* which are used when a set of related branches, which make some logical unit-of-change, are merged together. For example, two different parametrization schemes may be developed in their own branches but be related to each other because of some scientific interaction. In this case, a packaged branch may be used to manage a merging of these two related changes.
- *Configuration branches* which are similar, but are designed to be used to define a whole model configuration. This would typically be the result of merging packages and development branches.

Outside software changes, the configuration developer would benefit from tools to help with their need to keep a log of activity. This would make development more rapid. Provision of revision control for UMUI jobs will help in this regard, but there is a wider environment to consider. Most users and projects use wiki pages to manage this today, but a more integrated environment would be beneficial.

Another key benefit of the use of formal tools and well defined processes, including recording, is that they support changes in staff much better than manual and ad-hoc methods. However, ad-hoc methods applied by organised people do work reasonably well and unexpected changes in staff are rare. Hence, improvements in this area would be beneficial rather than essential, and it is hard to quantify the level of the benefit that would be obtained. Having well defined processes and supporting tools would likely lead to more rapid staff development as best practice would be more visible. Any solutions would almost certainly need to rely on bespoke development but would hopefully be based on available 'component technologies'.

4.5 Role 4. The Configuration Manager

The configurations of concern to the configuration manager should not be confused with a version of the software base. At the Met Office, a climate configuration, such as HadGEM3, can be selected and run at multiple UM versions, and a version of the UM can support multiple configurations, for example, for climate and weather prediction runs, for global and limited area, at high and low resolution etc.

As we have seen in the previous section, the benefits of integrated tools is an emerging need for the configuration developer, but this is recognised as even more beneficial for the role of the configuration manager who's job it is to document and act as guardian for the configurations defined by the configuration developer. In the absence of integrated tools, the configuration manager will use available methods for recording the information required to reproduce runs. Careful use of the UMUI, setting up new jobs as things change, supported by documentation on project wiki systems provide a simple, but effective solution but does involve some manual activity and personal diligence.

It should be recognised that configurations need to evolve over time. A configuration for a scientific aim may be created at a particular time but, as with software, through use, flaws will often be found and 'patches' may be required. Again, this process needs to be properly managed to provide clarity and avoid wasted effort and time.

The configuration manager, as with the configuration developer, would benefit from a set of tools that cover the full range of issues we have dealt with for the other roles. To summarise, the main configuration management issues are associated with:

- Source code
- Job options and parameters (i.e. UMUI settings and any 'hand edits'[6])
- Input files
- Ancillary files
- Platform used for the integration, including compiler version and the selected compiler options.

These issues all affect the results obtained from the integration. Other information, that does not affect the results, is also useful:

- The results obtained for a particular integration with a given configuration.

---

[6] Hand edits are typically provided by batch sed scripts to fill the gap resulting from the UMUI not being consistent with changes not yet on the trunk. The result of any hand-edit is archived with the job output.

• A record of known issues and problems with the configuration.

It is worth noting that the use of packages and configuration branches within the FCM system at the Met Office is equally targeted at supporting the configuration manager and the configuration developer.

4.6 Role 5. The Model Release Manager

This person is responsible for the management of the release of a new version of the code base. As we have pointed out already, in the case of the UM, a climate configuration is often available at multiple releases of the code base.

The model release manager's role is to ensure that proper governance is placed on the changes that make it into a new release. The UM is released every three or four months and a typical release involves about 80 branches (logical code changes) submitted by about 60 different developers. There are several hundreds of users of the system who will almost always need to run with a number of code changes (branches).

The model release manager must ensure that:

• code quality is good and coding standards are adhered to;
• the release can be completed on time to meet the needs of the customers of that release;
• supported scientific configurations are not impacted by changes targeted at other configurations;
• scientific and technical changes are put into a release in a consistent way. For example, technical changes required to support a scientific change are made available in time.

Typically, the scientific changes targeted at a release are ones relating to the requirements of key scientific configurations. The aim is to reduce the number of (FCM) branches used in order to simplify the configuration managers task and to reduce the risk of merge clashes in key runs.

Use of the Trac (Trac 2010)[7] system, which is well integrated with Subversion (and hence with FCM) has been a great benefit to the model release manager at the Met Office. Further, Trac allows for flexible tailoring to support particular processes. For example, at the Met Office the model release manager has requested to be able to monitor the status of a ticket through key stages, such as scientific review, code review, approval and commitment to the trunk.

The role of the model release manager is that of a project manager, and the use of Trac, and its ability to provide an integrated wiki has proven very useful in ensuring that information is widely available, in an organised way, for all contributors. Trac provides information in a way that is both descriptive (through text in wikis

---

[7] http://trac.edgewall.org/wiki/TracGuide

and in tickets) and precise (via access through the same interface to both code and documentation). In environments like the Met Office, such systems are of crucial importance since there are many contributors and beneficiaries.

The benefits of automated testing have proven to provide significant efficiencies within the process and a new testing system, the User Test Framework, has been developed to replace the previous, less comprehensive, test harness. As we move to a more modular system, such testing will be key to ensure that interfaces between components are properly managed at a technical level. More challenging is the need to encapsulate the interdependencies of scientific codes on each other in tests, which would lead to development of unit tests for scientific codes. Such unit testing would complement the system testing because, at the system level the number of combinations of path through the code are too many to allow full testing to be practical. Greater modularisation would aid in this regard, for example JULES will be developed in its own repository.

4.7 Collaborative Developer

The trend in Earth system modelling is towards increased collaboration of what is, after all, a global problem. Collaboration between groups and institutions focuses on sharing scientific models, configuration definitions and infrastructure. Further, collaboration includes the sharing of results of integrations and, importantly, sharing information about the runs that produced the results to enable them to be of use to the scientific, and wider, community. Collaboration poses a number of challenges.

Firstly, the code and environment needs to be portable to allow collaborators to pick up latest versions quickly, so that developments can be better synchronised. Code portability is quite mature, but infrastructure portability is an area identified for development. Configuration management and modularisation will both play a part in this to allow the infrastructure to adapt to different user environments (schedulers, file system layouts, archive systems, working practices etc).

Models are getting more complex with increasing numbers of component models, and this leads to the need to combine models developed by a range of sources into new ESMs. For example, at the Met Office, the ocean and sea ice models are primarily developed by other institutions. In this case, the source code for these models is managed in a separate repository and the preferred solution for exchanging data between the models is through the OASIS coupler. OASIS allows the model codes to be independent, with their interaction controlled by metadata.

At another level, model integration is achieved via the scripts that support the whole coupled model environment. As things become more complex, a methodology for allowing a wider range of combinations of base revisions of a range of models will be required. To support this, a compatibility matrix would be necessary, as well as a user interface system that can easily incorporate new and disparate models. However, to achieve this, better modularisation at the code level will be required. It should be noted that climate models are quite hard to modularise because of the complex, and changing, interfaces between the models. Difficult examples include implicit solvers that cross the boundaries of logical components, as found in implementations of boundary layers. Another example is the need for advection services to be provided to chemistry models with configurable numbers of chemical species.

Further, model components are now increasingly being developed under collaboration agreements, such as the UKCA project which is developing the next generation chemistry model that will be used as part of the UM. Achieving a single shared repository would be the most convenient solution for the project but would provide some challenges in terms of repository management and security, especially when it is considered that the code base is used in an operational and production context. The solution to this problem to date is to export and import between an internal and an external repository. This is a controllable solution, but with some overheads.

At the time of writing, this issue is being reviewed at the Met Office and it seems likely that multiple repositories will be used in future, and these will be synchronized at suitable points. Tools are being provided to support import and export for the synchronisation and for shared work in between. This mechanism will have overheads, but will allow the parties involved to provide the governance that each organisation needs.

4.8 Summary and Discussion

In this chapter, we have focused on the processes required rather than the tools used. It is likely that any modern configuration management tool could be used for the job of managing code, but the problem is much wider that this and hence the processes built around the tools are key.

As with other climate centres, The Met Office has built up experience in managing model source code and the wider definitions of configurations over a long period. FCM provides a good set of tools to control code development and configuration managers will use more manual methods to manage items, such as job definitions, ancillary files and the like. Hence, with care, configuration managers are able to reproduce results. However, facilities to help with the issue of wider configurations are limited and the difficulty of providing a user interface against a fluid code base remains a challenge. Inexperienced scientists are not always aware of the benefit to the wider community of configuration management tools. Hence, making the tools relevant to the users is an important aim of any initiative.

Organised information oils the wheels and Met Office staff use Trac to support this activity. In particular, in addition to being able to identify who changed a particular line of code and what the commit message was, if Trac is used properly then one can also link straight back to all the documentation which accompanied the change. This may include the original requirement, design notes, test results, etc. which can all be extremely valuable when trying to understand a change which is affecting simulation results.

This section has discussed the topic of configuration management of Earth system models from the perspective several roles in one large organisation, the Met Office. At this point in time, Earth system modelling is making serious progress towards becoming a truly international, collaborative effort. To a large extent, the success of this step depends upon the adoption and development of configuration management processes and support tools that scale beyond the confines of individual institutions. To achieve this is not a trivial matter as it would take a large effort to move groups onto common tools and, even more challenging, processes. Different groups use different tools and work in quite different ways today and the communities are focused on developing the science, not the technical working processes. An important lesson is that effective management of change is needed at all levels; from the scientific and infrastructure software at the heart of the modelling process all the way up to, and including, the processes and tools supporting configuration management itself.

References

Matthews D, Wilson G, Easterbrook S (2008) Configuration management for large-scale scientific computing at the UK Met Office. IEEE Comput Sci Eng 10(6):56–64 http://doi.ieeecomputersociety.org/10.1109/MCSE.2008.144

Pilato C, Collins-Sussman B, Fitzpatrick B (2008) Version control with subversion. http://svnbook.red-bean.com

Valcke S (2006) Oasis4 user guide (oasis4_0_2). Technical report TR/CMGC/06/74, CERFACS, 42 Avenue G. Coriolis, 31057 Toulouse Cedex 1, France

Chapter 5
Building Earth System Models

Stephanie Legutke

When the choices related to scientific and numerical aspects of an Earth system model (ESM) have been made, and when the programming language is chosen and the equations put into programming code, i.e. when the source code of the ESM is written, executable programs have to be generated which can be run on the computing platform available to the scientist. The generation of executables from the source code of an ESM is called the 'build process' in the ESM workflow (see Chap. 2). In the following paragraphs, requirements for build systems of ESMs are discussed. As an example the Standard Compile Environment (SCE) (Legutke et al. 2007) used at the German Climate Compute Center (DKRZ) is described. This system aims at supporting any ESM written in Fortran or C/C++. It is part of a more comprehensive system supporting ESM workflows including tools to harvest metadata that describe the build process.

5.1 Requirements

In the field of Information Technology (IT), 'building a target' is used in a wide sense.[1] A 'target' can be an executable program or a library archive, or anything for which a build rule can be written (e.g. an empty directory). The build process of ESMs usually embraces the build of many such 'targets' and may be very time consuming. It has to be kept in mind that the model development phase is a step in the ESM workflow where a scientist passes a lot of time. This step typically involves several rebuilds in a day. It is therefore desirable to accelerate the build process as much as possible.

---

[1] See http://en.wikipedia.org/wiki/Software_build

S. Legutke (✉)
German Climate Computing Centre (DKRZ), Hamburg, Germany
e-mail: legutke@dkrz.de

R. Ford et al., *Earth System Modelling – Volume 5,* SpringerBriefs in Earth
System Sciences, DOI: 10.1007/978-3-642-23932-8_5, © The Author(s) 2012

On the other hand, the different 'targets' which have to be compiled may depend on each other, and the order of compilation matters. There exists a high risk of erroneous compilations which is proportional to the complexity of the ESM source code. A 'forgotten' recompilation may not be detected by the compiler or linker, and the ESM may nevertheless produce reasonable results. This is the worst case of an erroneous build process.

During model development, a build system should not only support the scientists by providing the possibility of a rapid compilation of modified source code, but should also prevent mal-operations and give extensive and understandable messages to help analyze the situation if problems occur.

In addition, ESMs may need to be run on a large variety of platform architectures in order to harvest all computational and data storage resources that are available. The platforms range from single-processor workstations to all types of High Performance Computing (HPC) facilities including scalar, vector, shared or distributed memory machines, embedded in homogeneous or heterogeneous as well as Grid environments (Chaps. 3, 6, and 7). In order to enhance portability, ESM build systems should therefore support a large number of compilers for each of the programming languages used. Typically, the source code of ESMs is for the most part written in Fortran with some C or C++ code added. In particular, compilers provided for Fortran code on HPC platforms may cause programmers to write code that is especially suited to be optimized for the specific platform. This optimized source code does not necessarily comply with general standards which may be an obstacle to portability. The use of different compilers and different platforms, as well as the support of different sets of compiler options [e.g. to support debugging, to generate fast code, to check portability (see next paragraph)] should be easy and, if possible, transparent to the user.

5.2 Methods to Accelerate the Build

As mentioned above, the build of an ESM is a step in the ESM workflow where scientists may pass a lot of time, especially during the model development phase. It is therefore desirable to accelerate the process of ESM building in that phase. Some methods to achieve this are commented below.

Use a fast compiler. One move toward build-acceleration is to use a compiler which compiles quickly. We mention as an example the NAG[2] compiler which is not only fast but is also popular due to its capabilities to detect non-standard coding features, potential obstacles to portability (this topic is addressed in Sect. 5.1). Other compilers are configurable to only accept code complying with the American National Standards Institute (ANSI) standard. However, while restricting the source code to this standard during model development enhances portability, this does not necessarily lead to fast executables. Also, using a fast compiler in order to accelerate the build process does not necessarily mean that executables are generated that run

---

[2] see http://www.nag.com

efficiently on the target platform. In order to build fast executables the scientists may be advised or even constrained to use the platform vendor compiler. It remains the risk for the model developer that an executable, while being able to run and produce correct results when compiled with one compiler, does not necessarily run and produce correct results when compiled with a different compiler. This risk can be reduced by building the ESM operationally and frequently for all target platforms and run short test experiments (e.g. every night, after each modification of source code). [3]

Cross compilation. At some computing centres, in particular if the compute server is a vector machine optimized to process large arrays, compilation for the compute server is made possible on a different (scalar) machine at a faster speed. This process is called cross-compilation and can considerably increase the build speed.

Parallel builds. The build process can also be accelerated considerably by using multiple processors for the compilation. The 'make' utility e.g. (see below) supports parallel builds. Since the order of compilation of source code files matters, it is important to ensure that the correct order of compilation is guaranteed especially in parallel builds. During the development phase, when only a small number of files is changed between rebuilds, parallel builds are not so essential. It may have a larger effect in this phase if the build system ensures that only the minimum of recompilation is done (see below).

Pre-compiled software. Even when a new project with an ESM is started, it is normally not necessary to compile all source code that is part of the ESM. General-purpose libraries (NetCDF, MPI, etc.), used in many projects are often installed by the system administrators of the computing centre. This is not only done in order to relieve the users from the burden to install the libraries, but also in order to ensure that library versions are used which have been optimized for the architecture of the computing platform.

Other software, which is used only by a smaller group or project, may also be pre-compiled if the group or project members are not supposed to change it. This may be common project libraries, coupling software (see Vols. 2 and 3 of this series), entire components of the ESM, or parts of it not under development in a particular project.

The decision on what to pre-compile and what to leave under the control of the scientists has to be taken by the lead of the project or centre (Sect. 4.5). It may be necessary for example to guarantee binary reproducibility of an ESM experiment for many years. The decision may then be, to keep not only the compilation of the ESM under control but also that of the libraries involved. In that case the use of shared-object or dynamic libraries may not be a good choice unless the generation of those libraries is also controlled in the project as well as the side effects on other users.

Minimize the amount of recompilation. Recompiling the minimum is the method with the largest potential to speed up a rebuild process (and thereby model development) and can, in addition, be used on top of the others. In general, only the source code that has been changed since the last build has to be recompiled.

---

[3] Example with Buildbot (http://trac.buildbot.net/)

A straightforward recompilation of the minimum is inhibited by the many dependencies between the different pieces of software. A subroutine, which includes source code either by a Fortran or a preprocessor[4] instruction, has to be recompiled if included source code is changed. Also, a Fortran routine containing USE instructions has to be recompiled if any of the MODULEs it USEs has been changed. Thirdly, preprocessor instructions may be controlled by cpp flags passed to the compiler to allow conditional compilation of source code. Conditional compilation can be used for several purposes. Most frequent is the use of compiler- or architecture- dependent source code for optimization and portability reasons (see Vol. 2 of this series). This is not an issue in our context since these flags are usually not changed if neither the platform nor the compiler are changed. Preprocessor flags are also used to temporarily activate slow but verbose source code to help debugging or understanding a program. Some ESMs use conditional compilation to chose between different auxiliary software (I/O libraries, inter-component (or coupler) communication, etc.), between different parameterizations of physics, or to get prepared to run in specific combinations of components. A change of the status of preprocessor flags implies a change of compiler input and therefore requires recompilation.

It is essential for a build system to ensure that not only the changed source code files, but also all source code that depends on it, are automatically and quickly recompiled in the right order. Given that ESMs may be combinations of a number of component models which usually consist of hundreds or thousands of routines or source code files, and given that often the different components of an ESM are developed by different people or at different research centres and not with the same coding standard, it is not a trivial task to establish all dependencies and ensure that an ESM build process always runs correctly.

5.3 The 'Make' Utility

At the core of most build systems is the 'make' utility. It controls the order of compilation and decides what has to be redone in a repeated build.

Simply speaking, 'make' makes a target for which the build-rules are laid down in an input file, called 'Makefile'.

In addition to the rules to make a target, the 'Makefile' lists, for each target, the prerequisites, i.e. those targets that have to be built before. Building a single target can amount to the build of a large software system, e.g. an ESM, and a 'Makefile' can have a length of hundreds of lines.

Figure 5.1 displays an extract of such a 'Makefile'. It will cause the 'make' utility to do the following: the target 'prerequisite-2', which does not depend on anything, will be made first by calling first the command(s) of rule-1-to-make-prerequisite-2 and then those of rule-2-to-make-prerequisite-2. Next, the target 'prerequisite-1' can be made by running the command(s) of rule-to-make-prerequisite-1, since it only

---

[4] See e.g. http://gcc.gnu.org/onlinedocs/cpp/

```
.
.
target-x: prerequisite-1 prerequisite-2

prerequisite-1: prerequisite-2
    rule-to-make-prerequisite-1

prerequisite-2:
    rule-1-to-make-prerequisite-2
    rule-2-to-make-prerequisite-2
.
.
```

Fig. 5.1 Extract of a 'Makefile'

depends on 'prerequisite-2' which has just been remade. The build of target-x is then also accomplished since it just consists of making its two prerequisites.

In Earth system (ES) modelling, the primary targets are executable programs, binary libraries, and compiled source file binaries. Key rules are accordingly to link binaries into a program, to create a tar-formatted archive from a list of binaries, create a shared library from such an archive, or to compile a source code file. 'Makefile's frequently include other targets not directly related to the build process, such as generating documentation, or auxiliary targets, e.g. removing some files in order to enforce a rebuild. Because 'make' rebuilds a target if any of its prerequisites is newer than the target itself by evaluating the time stamps, it can be used to ensure that at each rebuild only the necessary is updated (provided the prerequisites are listed correctly).

One prerequisite for a binary created by the compiler is the source code file itself. In addition, any file included either by a preprocessor or by the compiler belongs to the prerequisites. Tools exist, and are normally part of ESM build systems, which generate the list of these prerequisites for compilation targets (*.o or *.mod files) in 'Makefile's.

A tool similar in its functionality to 'make' is the Ant build system. It is described in more detail in Chap. 6. In contrast to the 'make' tool is does not seem to be used for building ESMs.

A popular variant of 'make', at least in the Unix world, is the GNU[5] 'make' utility (Mecklenburg 2004).[6] Unfortunately, 'make' does not represent a standard, and portability problems may occur when porting a 'Makefile' to a different platform. A way out is to confine the content of the 'Makefile' to the (most) portable features. Another option is to create Makefile's for different target platforms (compile server) e.g. by the open source tools 'automake' (see below), 'CMake,'[7] or 'SCons.'[8]

---

[5] http://www.gnu.org

[6] See also http://www.gnu.org/software/make/manual

[7] http://www.cmake.org/cmake/project/News.html

[8] http://www.scons.org/

'Automake' is, together with other GNU 'Autotools' (Vaughan et al. 2000) a widely used build system for software packages written in C/C++. The Autotools include a number of integrated utilities which create a build system configured for the envisaged system with a high degree of automation, at least for POSIX. One of the advantages of the Autotools is that it relieves the user of the burden to maintain files with specifications for all platform the software is supposed to run on. The tools include a number of tests for the presence of features and provide flexibility for the software developer to add new tests. To the former belong e.g. the test for existence of system header files, the search for software paths (compilers, loader, or libraries), and tests for availability of compiler and loader options. As noted above, the 'automake' utility can be used to create portable Makefile's. The 'autoconf' tool creates a script, the 'configure' script, that configures the 'Makefile' to run on the target platform. And finally, 'libtool' can be used to build (shared) libraries. Additional utilities, such as 'aclocal', 'autoheader', or 'autoscan', help to write input files to the primary tools.

None of the above-mentioned tools, besides 'make', seem to be used in ESM build systems.[9] A reason may be that model developers generally are reluctant to use third party software. In addition, software packages coming with the Autotools are intended to be installed only once on the user's machine. Ease of installation has a high priority. In contrast, ESMs are continuously changed, implying that ease of inclusion and testing of new source code is the primary concern. More effort is typically needed to optimize an ESM for an HPC platform than to install it. Moreover, most of the tools are primarily developed to give C/C++ support rather than supporting Fortran, the language used for the bulk of ESM source code. As a consequence, these tools give only limited support for the creation of 'Makefile' dependencies needed by ESMs. Also, relying solely on the Autotools in a build system of an ESM using pre-compiler cpp flags bears the risk that the build system does not detect that a source file has to be recompiled when the status of a 'cpp' flag is changed but not the file using it. Last but not least, 'configure' tests that have to be run on the compute server (e.g. check the machine's endianness) can not be performed at build time with cross compilation which is often used in ESM builds (see above).

The above-mentioned arguments or others may be the reasons that ES modelers tend to use hand-written 'Makefiles' together with build systems tailored to the needs of their ESM and their way of working. In the next paragraph a system is described which aims at supporting the build of any ESM.

5.4 Example: The SCE Build System

The development of the Standard Compile Environment (SCE) build system (Legutke et al. 2007) was started in the PRISM project (see Vol. 1 of this series). Since the end of Partnership for Research Infrastructures in Earth System Modelling (PRISM), it has been maintained and further developed as the build part of the Integrating Model

---

[9] CMake is used for the visualization package VTK.

and Data Infrastructure (IMDI) ESM environment at the DKRZ.[10] In PRISM, the SCE was interfaced with the Supervisor Monitor Scheduler (SMS) Graphical User Interface (GUI) system used at ECMWF (see Chap. 6). This had some influence on its design. The compile scripts are assembled from pieces of script code specifically for the build of a particular component on a particular target platform. A new feature, which can be formulated independent of the components, automatically benefits all of them. Similar to the FRE system described in Sect. 2.2, the SCE uses 'meta scripts' to create korn-shell scripts (compile scripts) that configure Makefiles and call 'make' to build the ESM. In contrast to FRE, the SCE 'meta' scripts are korn-shell scripts themselves.

Compile scripts can be created on any platform since the SCE does not include any feature tests like the Autotools (see above). The SCE does not develop shared source code. However, the build of some common shared code is supported such as the OASIS3/4 coupling software (see Vol. 3 of this series) and some scientific libraries. The use of third-party software for script generation is restricted to the korn shell, the m4 macro processor,[11] and 'perl' (Schwartz et al. 2008). This considerably facilitates the installation of the SCE.

ESM developer centres are mostly reluctant to include source code suggested by someone from outside the owner community. However, updates of ESM source code included in the SCE become difficult when changes are not applied by both sides, the ESM developers and the SCE developers. Therefore, the SCE, claiming to be able to support any ESM, does not enforce coding conventions on the ESM source codes. In order to enable the use of all tools of the SCE, however, the ESM source code files have to be organized in a flat directory structure. A script is provided which transforms the file structure from its native form to the one required by the SCE and vice versa. Other 'requirements' are to be understood as suggestions for 'best practices' that help with the automated creation of Makefiles or that increase the speed of compilation. To these belongs the requirement that each source code file should contain no more than one single Fortran module, and that the file name should be the same as the name of the Fortran module it contains. A failure to comply with these 'best practices' may generate the need to add hand-written dependencies to the list of dependencies generated automatically.

The SCE supports modularity in terms of model components by providing per-component compile scripts. For each component, a binary archive library is created. The decision whether a software package shall be considered a component is up to the model developer and requires a certain degree of independence of the code from other component codes. A source code package can be a component at any granularity (atmosphere, cloud, diffusion parameterization).

The SCE provides tools which generate the prerequisite lists for the 'Makefile' compilation targets. These tools detect the include and USE prerequisites across an arbitrary number of directories, as well as nested dependencies (files included in an include-file, or USE instructions in an include-file) (see Sec. 5.2) by parsing the

---

[10] http://www.dkrz.de

[11] http://www.gnu.org/software/m4/manual/

ESM source code. A second tool creates lists of cpp-flag prerequisites for compilation targets. If the lists are included in the 'Makefile's, the compile process ensures that those files and only those which contain a certain cpp-flag are rebuilt if the status of that flag is changed. The SCE ensures that in a component build process, all libraries the component links to are checked, unless libraries have been specified for pre-compilation. This relates also to component libraries which are linked to the component being built. The SCE supports parallel builds in the sense described above i.e. it can run with multiple processors. Compiler output from a build is written to directories automatically tagged with the compiler name. This allows the use of different compilers in parallel on the same platform. In addition a configuration acronym can be passed to the system, which will be used to tag directories and executables. This allows SCE to work with differently configured component models (e.g. with/without MPI, stand-alone/coupled) at the same time.

Recently, the SCE was extended to harvest the metadata available at compile time. The scripts (again korn-shell) run with the compilation and gather information available during the build (e.g. the path to the source code repository and the revision number, compiler names and revisions, paths to libraries, and cpp-flags. The information is stored in Extensible Markup Language (XML)-formatted files and can, with the appropriate tools, be easily transferred into any other XML schema. The information that is gathered can be used, together with the corresponding information from the IMDI run environment, to rerun experiments.

Harvesting metadata during the build process instead of controlling the build process by the specification of XML formatted metadata as it is done by the FRE system (Sect. 2.2) has the advantage that information can be saved that is available only during the compilation e.g. compile times. The user can also be relieved from the burden of providing exact information e.g. on the compiler revisions or on the source code archiving/versioning system (Subversion etc.).

The SCE is modular by design itself. All of its tools are independent from each other but are inter-operable. This allows the scientists to run only those tools which are actually required to be run. For example, if no dependencies have been changed, the tools which generate the lists of dependencies for the 'Makefile's, need not to be run. If the description of the build process is not a requirement the generation of the XML files with the metadata can be omitted.

The SCE is used for all models supported by the Data Management Group (DM) at DKRZ. Site and OS specifications are available for all platforms ever required (e.g. Linux/i686 Workstation, Linux/ia64 clusters, IBM AIX/powerx, NEC SUPER-UX/SX, CRAY UNICOS, SGI IRIX64/MIPS, SUN SunOS/Sun4u, FUJITSU UXPV/VPP). It was used to build the IPCC/AR4 climate models ECHAM5/MPIOM of the MPI-M and ECHO-G of DKRZ as well as the regional climate model CLM used to downscale AR4 scenarios.[12] It is the build system used in the CMIP5/IPCC/AR5 experiments with the ESM 'MPI-ESM' of the Max Planck Institute for Meteorology containing the atmosphere component ECHAM6, the land surface JSBACH, the ocean component MPIOM with the marine bio-geo-chemistry module HAMOCC.

---

12 http://www.dkrz.de/Klimaforschung/konsortial/clm

The CORDEX [13] experiments with the RCM[14] CCLM[15] will also be run in this environment.

5.5 Discussion

We discussed issues arising from the compilation of big and complex source codes as can generally be assumed to be the case with ESMs.

The key problem is to minimize compilation time because in the ESM developing phase multiple builds are usually performed each day.

Not discussed was the question of how to choose compilers, loaders, and their command options. On HPC platforms the user is often constrained to use the compiler installed by the vendor. Since most ESMs are supposed to run on a variety of platforms to harvest available computing time it is, however, good practice to try as many compilers as possible. The build systems should thus support switching between available compilers. Testing many compilers also helps to write portable code (see Vol. 2 of this series).

Also not discussed was the question how to make use of compilers to generate fast executables. This may be in conflict with writing portable code. In that case, both optimized and portable code should be provided. Most compilers have options for automated optimization. Setting compilers and their options is a general issue for build systems since care has to be taken that they are specified consistently. The different build systems support a change of compiler or OS by providing the possibility to store the specifications in configuration files (FMS, FCM, see Chap. 4) or in environment parameters or in both (SCE). Most build systems try to insulate the users from these and other platform-specifics.

A number of methods have been described to accelerate the build process. In a situation where rebuilds are frequently done the most efficient one is to recompile only the minimum. Due to the many dependencies (executables on libraries, libraries on binaries, binaries on source code, source code on include files or on Fortran MODULES, etc.) it is not a trivial task to always guarantee that the right files are compiled in the right order. This is a critical issue since even with neglected dependencies the build may produce a running program and it can easily be overlooked that something has gone wrong.

A tool which supports the reduction of work in a rebuild and which is used by almost every group is the 'make' utility. The 'Makefile'(s), where the targets, the rules to build them, and the dependencies between them are laid down, are in most groups tailored to the particular needs of the ESM and the working practices of the

---

[13] COordinated Regional Downscaling EXperiments, http://wcrp.ipsl.jussieu.fr/SF_RCD_CORDEX.html

[14] Regional Climate Model.

[15] COSMO-Climate Limited-area Model, http://www.clm-community.eu

group. They may also depend on the structure of the source code, coding conventions, and the ways the different components can be assembled into a running ESM.

As an example, we described the SCE developed in the PRISM project. A design requirement was that it should fit to all component models in the project without requiring changes to the source code. It was designed to support a flexible exchange of components as well as a variety of platforms. Naturally, the design supports modularity in terms of components representing the ES. This is a feature the SCE has in common with other systems aiming at supporting a large number of component models in different combinations, e.g. FMS (Chap. 4)

While most ESM 'Makefile's are hand-written, there is normally support for automated generation of prerequisites for compiler output binaries. This is a consequence of the large number of files and the many dependencies in the source codes. It is straightforward as far as Fortran MODULEs or INCLUDEs are concerned, however, it becomes more tricky if changes of source code implied by preprocessing should accurately be taken into account. The build system of FCM, developed at the UK Met-Office which also claims to be a general-purpose build system, does this by comparing pre-processed files between builds and updating the dependencies for changed files. This system also reacts to changes of compiler options which is also not supported by 'make'. A specific feature of FCM is also that it creates a list of files to be compiled by analyzing subroutine calls. This is also a step towards minimizing recompilation and works well with routines contained in Fortran MODULEs, however, it may require some changes to the source codes for stand-alone routines.

An issue which becomes increasingly important is the question of metadata. Metadata describing the compilation process is important for a later re-generation of experiment results. They can be selected from the specifications which control the process, either set by the user or by the system. This is the line taken by the FMS system which is controlled by XML-formatted input. Acceptability may be a problem since the XML-format is not easily edited by humans directly. The development of GUIs which also check the validity of the user input will be an important contribution to improving such build systems. Another option would be to gather metadata from the scripts performing the compilation and from compiler and command output during compilation. This is the option the SCE has chosen.

References

Legutke S, Fast I, Gayler V (2007) The standard compile environment of IMDI. Technical report no. 4, Gruppe Modelle und Daten, Max-Planck-Institut für Meteorologie, Bundesstrasse 53, D-20146 Hamburg, Germany

Mecklenburg R (2004) Managing projects with GNU make, O'Reilly Media. ISBN:978-0-596-00610-5.

Vaughan G, Elliston B, Tromey T, Taylor I (2000) GNU autoconf, automake, and libtool, SAMS Publishing, Berkeley ISBN-10:1578701902

Schwartz R, Phoenix T, Foy B (2008) Learning perl, O'Reilly

Chapter 6
Running and Monitoring

Claes Larsson

6.1 Introduction

Job monitoring and management are the process of supporting the automation of running complex and large numbers of jobs, i.e. groups of commands in a file, on target hosts. One such job is the building process as described in Chap. 4 which should ideally be monitored as a self-contained part of the overall workflow (see Chap. 2) using the chosen tool. The job structure, parallelism and available hardware to run on should all have been selected during the configuration phase as described in Chap. 3 and it follows that the chosen tool needs to be flexible enough to accommodate diverse environments in this respect.

6.2 General Architectural Principles of Job Scheduling Systems

Most systems consist of a central server scheduler process that accepts job definitions from clients. The jobs are scheduled and started by the server by some mechanism, usually via a batch submission system on the target host. The scheduler communicates with all the jobs to receive status changes that can be monitored and that will affect future scheduling of jobs. Communication between the scheduler and its managed jobs is critical and must be robust. The scalability and performance of these systems, i.e. the number of jobs each scheduler can support, depends on the network activity needed to communicate status updates by the jobs to the scheduler, scheduler to monitoring client status communication and the processing needed for scheduling. This can vary substantially depending on the mix of jobs and the number of

C. Larsson (✉)
UREASON, Maidenhead, UK
e-mail: clarsson@ureason.com

monitoring clients. A critical feature of the scheduling system is its ability to restart from a known point or after a crash, facilitated by some form of a journaling[1] system.

6.3 Requirements

The requirements of a running and monitoring system will generally vary depending on whether the users are a small group with homogeneous hardware available or an institute with diverse platforms and administrative support personnel available and if the users target a grid based system or a single or cluster based computer system. A small user group will appreciate ease of setup and user friendliness whereas an institutional deployment requires more complex features that can be supported centrally. Following is a list of requirements that covers both areas of use:

- Graphical user interface and web interface availability.
  A graphical user interface (GUI) is an essential part of the monitoring tool as it has the ability to relay information on a very aggregated level as well as a detailed level and switching between the two can be done instantly. The use of colours and graphics can make it very easy to get an overview of job status and progress. A web interface makes it possible to do remote monitoring by wrapping the tools protocol in HTTP[2] which can pass through firewalls and so increases its usefulness. A web services interface makes it possible to call various functions of the scheduler/monitoring system, such as starting a job, programmatically and most important, remotely.
- Job management functionality.
  The tool should support starting, re-starting, stopping, pausing and the killing of individual jobs as well as collections of jobs. It should be possible to determine the status of the job visually by for example icon or colour. Easy viewing of the log of a job, changing the job scripts and restarting a modified job must be provided as these features help debugging and error correction.
- Alert and alarm message interface.
  Obviously the GUI monitoring tool should receive all alarms and alert the user to them either audibly or visually to allow the user to take the proper action. Support of a standard Message Service system or an Enterprise Service Bus makes it possible to relay alarms to a variety of devices and enables triggering external functions.
- Application Integration with configuration/submission tools.
  Unless the configuration, queueing and monitoring tools are all delivered by the same vendor, hand tailored system integration will be needed. The integration

---

[1] Journaling systems log changes to the system in order to be able to recover in the event of a failure. Thus work is not lost and processing can continue from a known state.

[2] Hypertext Transfer Protocol (HTTP) is a protocol used by the World Wide Web for requesting resources via Uniform Resource Locators (URL), such as links in a web page.

process benefits from standardised XML[3] files that minimises syntactical errors and transports with ease.

- Platform support and installation requirements and level of expertise needed. Support for supercomputing platforms is important for these applications and this can be achieved via virtual machines such as Java or by source packages that can be compiled on most UNIX flavour systems. The level of expertise involved in compiling, running and maintaining the system is necessarily high as the complexity of these systems is high. However, users should be protected against many of the complexities by a GUI that gives access to information without the need for the user to have detailed knowledge of for example file locations.
- Configurability and adaptability. Special features suitable for climate modelling such as time stepping functions, disk usage etc should be provided or the chosen tool needs to provide means of extending or adapting to the computing platform and job environment.
- Historian and data mining. The job monitoring system has the capability of generating large amounts of data specific to the operational characteristics of the site which can be analysed in order to improve the throughput and the understanding of jobs dependencies. There is a close relationship between system monitoring and job monitoring in this respect but in this chapter we are mainly concerned with code performance and throughput.

This section has outlined a number of general requirements and we will now go on to look into detail how these requirements are supported in existing tools.

6.4 Tools Available at Major Climate Research Centers

6.4.1 SMS

Supervisor monitor scheduler (SMS)[4] and Command and display program (CDP) with its X-Window/Motif graphical user interface (GUI) front-end XCdp (see Fig. 6.1), is a product of the European Centre for Medium-Range Weather Forecasts (ECMWF).[5]

SMS is a C program that runs as a server in user mode communicating with clients via SUN RPC[6] using TCP/IP.[7] It is capable of servicing a large number of clients and has journaling facilities for recovering from failures. Many SMS schedulers can

---

[3] Extensible Markup Language, a system for organising and labelling elements of a document in a customised and extensible way.

[4] http://www.ecmwf.int/publications/manuals/sms/downloads/sms 2009.pdf.

[5] Many thanks to Axel Bonet at ECMWF for help with reviewing this chapter.

[6] Remote procedure call (RPC) is an inter-process communication that allows a computer program to cause a subroutine or procedure to execute on another computer.

[7] Transfer Control Protocol (TCP) / Internet Protocol (IP), TCP provides a communication service at an intermediate level between an application program and the IP.

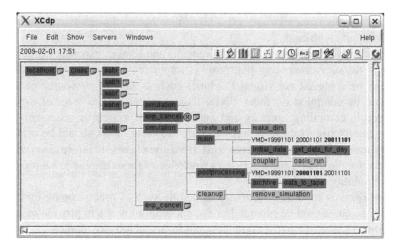

Fig. 6.1 XCdp job monitoring and visualisation window

Fig. 6.2 SMS task definition

```
# make_dirs.sms
%include <trap.h>
%include ''setup.h''
for dds in %ROOTDIR% ; do
    mkdir -p ${dds}
done
%include <endt.h>
```

run on the same host or in the same local area network (LAN) using different RPC ports and names. The GUI clients "discovers" the servers using a UDP[8] broadcast on the LAN. CDP command line clients log in to the SMS server to control and monitor jobs. SMS defines tasks, families, suites (which are sets of tasks), see Listing 1, via its SMS definition language in files called scripts (see Fig. 6.2). Tasks definitions/files are submitted (played in SMS terminology, as seen in Fig. 6.3 number 1.) via CDP to SMS for scheduling and are assigned a status (queued, submitted, active, aborted, completed etc). Various events, such as task completion, meter progress, user initiated killing etc, are reported to SMS and cause the tasks status transitions.

Interaction with SMS is through CDP, the SMS command language interpreter (CLI). It is text based or GUI based with XCdp (see Fig. 6.1). The user starts CDP and can then access all its commands to display and manage his tasks.

---

[8]   User Datagram Protocol, an Internet Protocol for sending messages, in this case referred to as datagrams, to other hosts on an Internet Protocol network.

6.4.2 Installation and Hardware Support

SMS is UNIX only software used on a variety of UNIX systems and portability is generally good for UNIX flavours. It is installed using make and configure which allows the user to configure important parameters such as default submission method. However, dependencies on X, Motif and C libraries requires more effort to rectify should problems be encountered compared to for instance a system written in Java only.

6.4.3 Task Definition Language

A UNIX-style shell scripting language influenced by the C macro preprocessing language is used for the tasks script. The task skeleton, a plain text file, will contain SMS-variables and include file directives to be replaced or expanded when SMS generates the actual instance of the job file for its submission.

Tasks are referred to hierarchically like a UNIX file system, and the corresponding job script will be created in a directory accordingly. This creates a homogeneous way to structure the definitions. The combination of these traditional and powerful UNIX concepts allows the user to define and configure almost any type of task and provides a welcoming environment for the UNIX affectionado.

Jobs are structured into suites, see Listing 1, that contains one or more families. Each family is made up of tasks (see Fig. 6.2). The tasks contain the real work being done, for example, make_dirs.sms in family create_setup.

The include files will be pulled in by SMS's preprocessor just like in a C program, where angle brackets stands for system file and quotes for "user" files, typically containing definition of variables as seen in Fig. 6.3 number 5. Preprocessing allows for layering of functionality and ease of maintenance, for example endt.h contains "end task" processing such as running a program to notify SMS that the task is completed. This functionality stays "hidden" and does not detract attention from the main functionality of the make_dirs task.

6.4.4 Macro Variables

CDP works with three kinds of variables:

- Shell variables imported from the environment using the 'setenv -i' command. referred to as $VARIABLENAME.
- CDP variables defined in the script with the 'set' command. Also referred to by $VARIABLENAME, as seen in Fig. 6.3 suite.def.

- SMS variables used as %VARIABLENAME% in the task script, such as %SMSHOME% and defined with the edit command, e.g. edit ROOTDIR "$ROOTLIST", in the definition file (suite.def). Some of these variables are generated automatically by SMS, e.g. %SMSJOB% which defines the UNIX path where to access the job related to a task.

Variables are defined hierarchically and inherited from outer layers. In Fig. 6.3 task make_dirs.sms inherits MEM = 100 from the outer suite.def. The preprocessed task becomes a job as in make_dirs.job and sets the value to MEM = 100, the inherited value, for its execution.

6.4.5 Dependency Declaration

Explicit dependencies can be made on a time, date, another node status, meter or event. Dependency declarations are described using 'triggers' that work with statuses, such as "task complete", referring to the triggering task with its relative or absolute path within the suite, or from another suite's task on the same SMS server. Thus triggers define the sequence of processing of the tasks. Triggers are usually defined for SMS nodes, such as tasks in other families but can be any logical expression, e.g. the value of a variable (datalen):

```
trigger../anotherfamily/init == completeand../init : datalen > 10
```

Special display features like "meter", "label" etc exists to visualise variables and states as seen in Fig. 6.1 for the variable YMD.

6.4.6 Starting Tasks

In typical SMS fashion jobs are started with a macro command SMSCMD, which itself can contain macros such as %HOST%, %SMSJOB% and %SMSJOBOUT% which is the name of the output file. An example command would be:

```
edit SMSCMD ssh %REMOTEHOST% %SMSJOB% 2>&1> %SMSJOBOUT%
```

which amounts to ssh being executed with the translated values of the %macros%, as seen in Fig. 6.3 number 2. Output is sent to %SMSJOBOUT% (no. 4) and the resulting job file %SMSJOB% (no.3) is executed.

6.4.7 Special Large Installation Features

File viewing on remote systems can be handled automatically via a logserver. The smsfetch command allows for integration of source control systems, WebDAV[9] etc. by defining a command to access a file and read the output stream of that command. A "limit" command provides basic load management by limiting the number of tasks to sms executing simultaneously in a suite based on the value of limit. Documentation can be placed in task files and viewed via XCdp, making it easy to access when needed.

6.4.8 A Use Case

We will now define a suite, as seen in Listing 1, which consists of a 'simulation' family split into a setup, main with initial data, a coupled run and an archiving task. The suite will be named after the owner, have an id and run on $WSHOST. These variables are set in the environment and imported to SMS with the 'edit' command.

   We prefer that

- The processing is made for a sequence of days, line 13, of type date which makes logical operations such as "less than" meaningful. We define a function which will insert the repeat command at the right place to allow for iteration of the tasks, "repeat_command".
- The main tasks run when setup is complete and do not run ahead of the post-processing, line 21.
- The post-processing of a day only starts after the main run for that day, line 34.
- The cleanup finishes when the post-processing (and hence main run) is complete, line 41.

   We define a start command SMSCMD that calls the script "smssubmit.x" to start off the simulation, line 10.

   The structure of the suite can be seen with XCdp while executing (see Fig. 6.1) where the current day main tasks can be seen executing with the value of variable YMD shown.

6.5 XCdp GUI

XCdp is the GUI front-end for CDP. It allows access to most CDP commands using configurable menus accessible with the mouse-pointer.

---

[9] Web-based Distributed Authoring and Versioning, or WebDAV, is a set of extensions to the Hypertext Transfer Protocol (HTTP) that allows users to access files (resources) on web servers.

Listing 1 Suite definition for our use case

```
 1   suite $OWNER
 2
 3   # from prepIFS
 4
 5     family $EXPVER
 6       edit EXPVER      $EXPVER
 7       edit USER        $USER
 8       edit WSHOST      $WSHOST
 9       edit OWNER       $OWNER
10       edit SMSCMD      'smssubmit.x %SMSJOB% %WSHOST% %
             USER%'
11
12   define repeat_cmd { # Called for every stream of
         families
13     repeat date YMD 19991101 20001101 20011101
14   }
15
16     family simulation
17       family create_setup
18         task make_dirs
19       endfamily
20       family main
21         trigger create_setup==complete and ( lag:YMD
             le main:YMD )
22         smsmeter $YMD
23         repeat_cmd
24         family initial_data
25           task get_data_for_day
26         endfamily
27         family coupler
28           trigger initial_data==complete
29           task oasis_run
30         endfamily
31       endfamily
32       family postprocessing
33         repeat_cmd
34         trigger main:YMD gt postprocessing:YMD
35         family archive
36           task data_to_tape
37         endfamily
38       endfamily
39
40       family cleanup
41         trigger postprocessing==complete
42         task remove_simulation
43       endfamily
44     endfamily
45    endfamily
46   endsuite
```

6.5.1 Main Monitoring View

The main monitoring new (see Fig. 6.1), showing the suite defined in Listing 1 and its families and tasks colour coded for state information: green for active, red for aborted, yellow for completed etc. Nodes can be collapsed and expanded by a click. Each node hosts a menu with the related functions to manage the node's status such as Execute, Re-queue, Edit etc. The colour coding together with the aggregation of state and collapsible hierarchies of nodes makes it very easy to monitor and manage very large trees of tasks.

6.5.2 Task View

When a task has failed, indicated by it's node turning red, the user needs to understand why and also be able to rectify the problem. Using the output function on a node in the main view opens the task view (Fig. 6.4). This view allows the user to look at the output, edit the task script and the defined variables (Fig. 6.5) without resorting to console commands and file locations. Once the problem has been found and fixed XCdp allows the user to immediately execute the edited job from a menu on the node.

A number of subviews shows the tasks timeline, triggers (dependencies) and documentation, providing information at the click of a mouse button (not shown here).

6.5.3 Security

Security is rudimentary with either userid/password or userid with distinction between read-only and full access users. Since SMS will normally run in a LAN this is quite sufficient.

6.6 Ant and Frameworks

Another Neat Tool (Ant)[10] is a build system similar to make. It uses property files for macro name initialisation and is XML based. Ant tries to build tasks specified in the XML file using built in commands such as copy and makedir (see example in Fig. 6.6). It is implemented in Java and is easily extended with new functionality through the addition of user defined tasks implemented as Java jar files in a standardised way. Although it is primarily a build system it can execute any task including running

---

[10] http://ant.apache.org/.

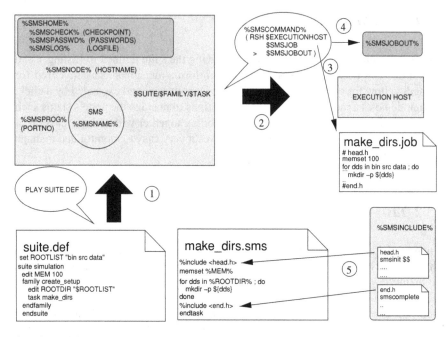

Fig. 6.3 SMS files and generation process process

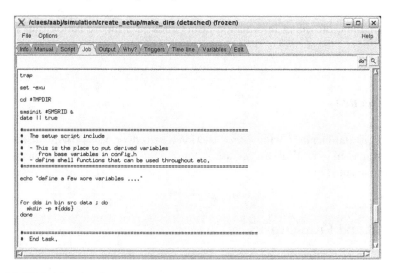

Fig. 6.4 XCdp job script view

a program (see example in Fig. 6.7) for such a task. Frameworks have been built around Ant to allow for running multiple tasks on multiple machines, for example TeamCity.

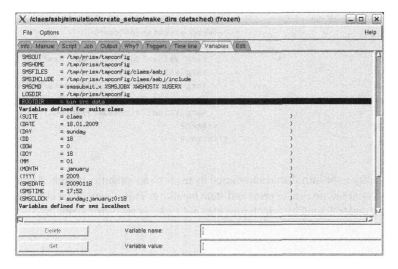

Fig. 6.5 XCdp job variables view

Fig. 6.6 Example of run
target in Ant

```
<property name="version" value="${basedir}/2.1"/>

<target name="prepare">
    <mkdir dir="${version}"/data>
    <copy todir="${version}/data">
        <fileset dir="${env.DATAHOME}">
            <exclude name="meta/**"/>
            <include name="*.dat"/>
        </fileset>
    </copy>
</target>
```

TeamCity is a Java-based build management and continuous integration server from JetBrains.[11] TeamCity features a server that communicates with a number of agents that execute on remote platforms and run the tasks.

A web interface makes it easy to follow job progression but it lacks advanced job control. It is strongly geared towards the building process of complex software and as such would make an excellent tool for Earth system modelers looking for a build environment rather than a job scheduler. Coming back to Ant we should note how easy tasks are specified using macro and built-in commands together with an extensive file selecting utility. In the example in Fig. 6.6 the task prepare is defined.

Using the macro name version the version and data directory is created and all data files are copied from DATAHOME to data. The run-model task (see Fig. 6.7) is made dependent on the prepare task with the depends property. The use of macros, file-sets and built-in commands facilitate a structured and flexible task hierarchy that sets the standard for task specification languages. Ant's extension facility allows new

---

[11] http://www.jetbrains.com/.

Fig. 6.7 Example of run
target in Ant

```
<target name="run-model" depends="prepare">
  <exec executable="cmd">
    <arg value="/c"/>
    <arg value="fortran model.exe"/>
    <arg value="-i"/>
    <arg value="${basedir}/indata/${DATE}"/>
    <arg value="-o"/>
    <arg value="${basedir}/output/"/>
  </exec>
</target>
```

functionality to be introduced and shared by users so adaptability is high. Ant's weak point is that it has no proper error and state handling. This is somewhat compensated for if run under TeamCity which provides notification and error signaling.

6.7 Job Scheduler

An alternative to SMS is Jobscheduler[12] from SOS GmbH. It is an Open Source Java based and GPL[13] licensed system but a commercial licence will buy you support and the possibility to ship the product with your own software with a non GPL licence. It is available for most UNIX flavours and also Windows. The architecture differs from SMS in that the scheduler runs on each machine where jobs execute. A central database contains the scheduled jobs. This adds complexity to the installation. On remote execution the remote Job Scheduler functions as a slave. Failover systems can be configured for high availability. Task chains are defined by adding jobs to them with state information. An "order" token is passed between the jobs, also called nodes, which is a datastructure holding state and other information. See Fig. 6.8 which shows a JavaScript[14] job definition. Here a chain "JobChain:" is created. Jobs "job_100" and "job_200" are added with input state, output state and error state defining the progression, error and completion states. Jobs are defined in XML files and management functions are accessible via the command line, a GUI or an API with multilanguage bindings such as JavaScript. The system has the expected features of starting jobs in sequences, at a particular time etc as well as job prioritisation. Error handling is similar to VMS[15] scripting with "on error" actions and is different from SMS. It also features "drop" directories which are watched by the system for changes. Tasks copied to these directories become "known" instantly to the system. Tasks can be exposed as Web Services using an XML API that allows for

---

[12] http://www.sos-berlin.com.

[13] GNU General Public License free software license. The GPL is the first license for general use, it is intended to guarantee your freedom to share and change all versions of a program and to make sure it remains free software for all its users.

[14] JavaScript is a scripting language used to enable programmatic access to objects within other applications.

[15] Virtual Memory System, Operating system by Digital Equipment Corporation.

Fig. 6.8 Example Job
Scheduler job chain
Javascript definition

```
var my_job_chain = spooler.create_job_chain();
my_job_chain.name = "JobChain";

my_job_chain.add_job( "job_100", 100,  200, 999 );
my_job_chain.add_job( "job_200", 200, 1000, 999 );
my_job_chain.add_end_state(  999 );
my_job_chain.add_end_state( 1000 );
spooler.add_job_chain( my_job_chain );
```

Fig. 6.9 Example Job
Scheduler Sh script

```
<job name="simple_shell">
  <script language="shell"><![CDATA[
    echo hello world
    call my_script.cmd
    my_prog.exe
  ]]></script>
</job>
```

Fig. 6.10 Example Job
Scheduler include statement

```
<job name="simple_command">
  <script language="shell">
    <include file="my_script.cmd"/>
  </script>
</job>
```

control of the tasks. Graphical User Interfaces are provided for starting and stopping jobs as well as monitoring with the Operations GUI which is web based. A GUI Job Scheduler Editor allows for point and click definition of tasks which are saved as XML.

From the examples (taken from Job Schedulers documentation) in Figs. 6.8, 6.9 and 6.10 we can see that the style follows the Ant tradition and have the usual verbosity expected in XML files.

6.8 Conclusion

The best features of SMS are the user interface and the job management/control, as is evident from the presented figures.

The ability to make functions for time steps etc makes it possible to tailor boilerplate templates suitable for climate modelling. It is very robust and scalability is excellent, many SMS systems can run on the the same host or in the same environment. It is very flexible and can be adapted and extended to work with any queueing system or tool. This strength can also be seen as a weakness as there is no out of the box functionality.

The initial investment needed to start using SMS makes the system ideal for institutional use where new users can get support and will not have to tackle integration and installation issues.

In comparison, a Job Scheduler has a different architecture, offers Web Services integration and the use of XML based configuration files with a different error handling system and can run on Windows. Both systems are very capable but very different in style. Apart from the Web Services functionality, which will be an issue in situations where the modellers are not all working at the same site, this is the major point. In a UNIX only environment SMS will fit in due to its design philosophy and the user will work in a natural and intuitive way with the system. In an environment where the modelling system is built and run on XML based knowledge structures Job Scheduler will be the choice.

6.9 Future Directions

As we become connected to the Internet 24/7 everywhere, future systems need to integrate with the Internet seamlessly to allow whatever hardware client you are using, for example a mobile phone, to be used for monitoring. Operations data mining metrics, that is generating and collecting operational statistics on jobs and using these to find patterns for fine tuning the scheduling system, and new approaches to scheduling problems borrowing from genetic algorithms and other evolutionary models will result in autonomic self adjusting schedulers with multiple goals for optimisation on data locality, access and replication, throughput, resource utilisation, parallelisation etc. These areas are researched in virtual server and Grid/Cloud computing environments and eventually results will also optimise non distributed scheduling. For climate modelling which accesses large amounts of data the coordination and scheduling between compute grid systems and data grid systems may make it possible to run very large simulations efficiently. Visualisation of complex job loads will be linked to data mining to allow for predictive monitoring and manual adjustment of goals.

Chapter 7
Configuring, Building and Running Models in GENIE

Gethin Williams

The following casestudy provides a workflow perspective on the Grid ENabled Integrated Earth system modelling (GENIE) framework, which was introduced in Vol. 1. The three aims in constructing the framework were; (i) the flexible coupling of Earth system components; (ii) to tune and run the resulting models on a wide variety of platforms, including the Grid; and (iii) to archive, query and retrieve the results.

The component models incorporated into GENIE to date have been designed to be computationally inexpensive. This feature allows researchers using the framework to construct ensemble experiments to investigate long-term climatic trends or to conduct comprehensive parameter-space or climate sensitivity studies. GENIE also incorporates a full carbon cycle which makes it particularly suitable for recreating paleo-climates spanning the full ice-core record. The framework also has a role in simulating future climate change, including the response of slowly varying features of the Earth system, such as the deep ocean, sediment deposition and mineral weathering processes.

The workflow description is split into four main areas. First is the task of gaining access to the model files, including source code, documentation, configuration descriptions and sample input and outputs. When considering access to files, it is natural to look at the facilities provided by version control systems (VCSs). Section 7.2 looks in more detail at the categories of information that must be recorded in a model configuration. The section also reflects upon aspects such as the ease of use of model configurations as well as their provenance and evolution when harnessed to a growing model. The third area is the task of building an executable. Topics of particular relevance to the GENIE framework in this regard include code portability and dependencies as well as the incorporation of build-time model configurations. Lastly, we will outline some issues relevant to running and monitoring the model in the setting of a large ensemble or tuning experiment.

G. Williams (✉)
School of Geographical Sciences, Bristol University, Bristol, UK
e-mail: gethin.williams@bristol.ac.uk

R. Ford et al., *Earth System Modelling – Volume 5,* SpringerBriefs in Earth
System Sciences, DOI: 10.1007/978-3-642-23932-8_7, © The Author(s) 2012

7.1 Access to the Modelling Framework

The first step in the workflow is to gain access to the source code. The GENIE framework contains component models developed by researchers from a number of institutions and has users situated in many more locations around the globe. A single point of contact through which to marshall all this disparate activity is essential for the enterprise. Users and developers alike require access to model files with unambiguous provenance and access permissions must also be appropriate to all the interested parties. Fortunately VCSs are designed to meet this very need. For GENIE, we adopted a mature and popular open source VCS called Subversion (Pilato et al. 2008).

In addition to providing a single point of access for the code, Subversion offers many other features which help to organise development effort. The first is fine control over access permissions. For example, it is a straightforward matter to give a user group read-only access to the files. Developers are then free to safely release the model files to a cohort of scientists who use and test the framework, feeding back bug reports and development requests. Source code editing rights can be further refined such that separate groups of developers may—if they wish—retain sovereignty over particular component models in the framework. (Subversion allows specification of access control right down to the individual user and file level). This can give different development groups piece of mind and can help prevent the introduction of bugs through inadvertent edits to the code-base. The job of oversight and control of access falls to the roles of configuration developer and manager, as identified in Chap. 4.

It is to be expected that scientists from different institutions will have different research agendas. Sometimes these agendas will overlap, but at other times they will be at odds. A feature of version control which helps to marshal these occasionally competing efforts is the ability to *branch* a development line from the main *trunk* [see (Pilato et al. 2008) for more explanation of this terminology]. A development branch can accept any number of modifications without disrupting the trunk or, indeed, any other branches which exist. However, as a branch remains inside the same repository of files as any other development line, any successful innovations can be easily merged from the branch back into the trunk. (Subversion has a number of features to automate and assist the merging process). Should they not prove useful, branches can be simply discontinued.

The topic of the scientific cornerstone of reproducibility has arisen in a number of the previous chapters. Key to this goal is the ability to re-run an experiment using the exact same source code as was used the first time. A VCS preserves access to any previous revision of a model and so is an obvious boon to a scientist looking to replicate an experimental result. Using Subversion, a particular—and well scrutinised—version of the code base can be *tagged* with a symbolic release name (e.g. rel-x-y-z). All experiments should be carried out using a released version of the model, where future access to that version is greatly simplified by reference to it's symbolic name.

An operational model comprises more than just the source code. Other key elements include reference input and output files, configuration files and documentation. We will discuss reference and configuration files in forthcoming sections. Two reasons for keeping documentation files in the same repository as the source code are so that; (i) they can be distributed to users together; and (ii) the documentation is more likely to be maintained if it is stored alongside the code that it describes. Automatic documentation generators can be particularly useful in this regard, as they create well formatted, easily digestible documentation from appropriately commented source code. In GENIE, we have adopted Doxygen[1] for this task, as it is a widely used, open-source tool which has recently expanded its range of supported languages to include Fortran. GENIE also employs a project wiki to provide a quick and easy access to sometimes ephemeral information about the framework. Since web-based wiki systems are easily edited, they can provide a vibrant forum through which distributed developers can coordinate themselves. This dynamic facility is distinct to the longer lasting, more polished material contained within a project manual. Chapter 4 highlighted the Trac project management and bug/issue tracking tool (Trac)[2] which closely couples wiki-based management of bug reports, tickets and road maps with a VCS. If the user or developer cohort were to expand, GENIE would move to using a coordination tool such as Trac.

7.2 Model Configuration

The importance of configuration management and its place in the overall workflow of assembling and running an Earth system model (ESM) was highlighted in Chaps. 4 and 2 respectively. Both those chapters, together with Chap. 5, make reference to the challenge of reproducibility. A standardised workflow and access to well managed configurations both have an important part to play in meeting that challenge. The particularly flexible nature of the GENIE framework makes configuration management an especially important topic for model developers.

In common with many other projects, the GENIE modelers have chosen the Extensible Markup Language (XML) as the format for storing the model configuration. XML has a number of attractive properties for this task. The first is highlighted by the word 'extensible'. An evolving model will (hopefully!) incorporate new science—perhaps whole new component models—over time and so will require additional settings to control the enhanced features. To remain useful, any means of model configuration must keep pace with model development. Since new markup tags can be created and added to XML documents, new structures can be incorporated while still retaining those already in place. Other attractive properties are that the format is machine readable and that many mature tools exist for parsing and validating XML documents. This means that model configuration files can be conveniently

---

[1] http://www.doxygen.org

[2] http://trac.edgewall.org/wiki/TracGuide

exploited by larger (and potentially automated) workflows, such as a search over parameter-space using a model ensemble, or—more prosaically—running a suite of model regression tests. A downside to using XML, as identified in Chap. 5, is that the format is not easily read and modified by users directly, due to the superabundance of—often verbose—tags. The use of a GUI to present a user-friendly view of the configuration is a tempting solution to this problem, but is yet another component in the workflow to create and keep up-to-date.

Some of the partial differential equations used in ESMs are highly sensitive to initial conditions and as a consequence the output values can be highly dependent upon elements of the computing environment, such as third party libraries, compiler version and flags, operating system and central processing unit (CPU) architecture. Accordingly, these details must be recorded in the model configuration if bit reproducible results are required for certain ESMs.

In addition to providing access to (hopefully well scrutinised) model configuration files of known provenance, version control has much to offer configuration management. For example, model configurations are often the result of a model tuning exercise and, using version control, it is possible to store the chosen configuration at the same revision as the model files used to derive it. Another benefit of using a VCS is that the repository can be arranged such that a single checkout contains all the files needed to perform a validation test of the model. In GENIE, a testing harness is integrated with the build system through the specification of testing rules in the makefiles. These are stored, together with essential input and reference output files in the framework's repository. A single model checkout is self-sufficient and a suite of tests can be run immediately to determine whether or not the model is functioning as expected in its current context.

Parameter and several build settings for the model are sourced from a configuration file. This process is carried out using a mixture of shell scripts, makefiles and Extensible Stylesheet Language Translations (XSLTs) of the XML files. In order to present a user with as succinct a configuration file as possible (and in common with the workflow described in Chap. 2) *inheritance* is exploited and a configuration is described as a minimal set of perturbations to a complete set of defaults. Modification of the defaults is effected by merging the two (or more, if the inheritance line is deeper) XML files using XSLT. Further XSLTs are used to map entries from the resultant master configuration file into other formats. Some transforms extract compile-time model settings (e.g. grid spacing or the number of ocean tracers) and convert them into makefile format for inclusion by the build system. Others convert run-time parameter settings into Fortran namelist files, ready to be consumed by a running executable.

Storing configuration information together with source code under version control, integrating the process of model configuration with the build system and the use of an extensible, self documenting and machine readable file format have all eased the task of model configuration. A touchstone for the utility of model description in GENIE is the fact that a user need only type *make test* in order to compile and run a suite of regression tests in which outputs from reference configurations are automatically compared to previously stored and validated outputs.

However, this does not mean that the area of model configuration is without problems. A clear issue is user ambivalence toward XML. While the format has much to offer, it is tortuous for humans to read. This problem prompts the creation of tools to manipulate the configuration files, spawning yet more coding, testing, interoperability and maintenance headaches. One lesson to be drawn from the experience of constructing the above edifice is that the task of recording and applying settings would be a deal deal simpler if they were all given to the model at run-time and none specified at build-time.

7.3 Building

The second aim of the GENIE framework, as listed in Vol. 1, is to tune and run ESMs on a wide variety of platforms. This is consistent with the increasing trend toward collaborative model development across many institutions, often situated in different countries, as noted in Chap. 4. While this trend is laudable and consistent with a mature and robust model, it does present the software infrastructure developer (a role identified in Chap. 4) with challenges.

In Chap. 5 the topic of build system portability was raised, along with the need to embed metadata pertaining to the build (e.g. compiler flags) into the workflow. Section 7.2 described the XML configuration files used in the GENIE framework and how they can accommodate such metadata. Indeed, once configuration files are read by the build system, it is possible to use a conditional so that only the component models selected in the configuration are compiled and linked to form the model executable. Similar to the SCE system described in Chap. 5, the GENIE framework has adopted GNU make as the backbone of the build process. In principle, relying upon a particular implementation of the make utility (with some unique enhancements such as conditional execution and text manipulation functions) could limit portability. In practice, however, we did not find the use of GNU make to be at all limiting. This is quite likely due to the inexorable proliferation of Linux, appearing as it does today as the operating system of choice on many compute servers and clusters. We have also been able to build on Windows platforms with relative ease, through the use of the Cygwin[3] emulation environment. Another reason to favour writing makefiles is that we have been wary of the investment in developer time required to create a comprehensive and robust set of platform recipes for the GNU autotools approach to build system configuration.

Given the above approach, a key objective for any build system developer is to create a single, overarching set of makefiles which can be invoked very simply by a user. Our approach to this in the GENIE framework borrows the concept of inheritance discussed in relation to model configuration. Rules and default build settings (such as the set of flags chosen for use with a particular compiler and some preprocessor macros) are held in a large, typically unseen makefile.

---

[3] http://www.cygwin.com

User specific settings (such as the choice of compiler or the location of the NetCDF libraries) are made in a smaller, more palatable makefile. This user-facing makefile, along with another translated from the XML configuration file, are included by the large, static makefile and influence the build through overriding the defaults.

A recurring challenge to the project has been to ensure that the source code compiles correctly using many different compilers. Compiler vendors make different interpretations of the various Fortran standards (Fortran is the language of choice for a great many ESM developers), and so while code may be considered legal by one compiler, the same code may trigger errors with another. One way the project has addressed this has been to instigate an automated nightly build. Once a comprehensive build system is in place, and held with the source code in a version control repository, it is a simple matter to set up a scheduled task (e.g. a 'cron job' in Linux) which will checkout a fresh copy of the latest source code and attempt to build and exercise the model in as many and varied ways as the developers can conjure. Chapter 5 made reference to the CMake build tool.[4] The complementary tools, CTest (bundled with CMake) and CDash[5] exist, respectively, to promote regular testing over distributed platforms and the collation of the results on a single web page. The future adoption of tools like these would make a significant contribution to the GENIE project.

The use of an automated nightly build within a development workflow helps to quickly identify bugs in a code base, where the introduction of bugs is almost inevitable during any code writing. It is well established in the field of software engineering that a bug found early will take significantly less effort to address than one found later in the development cycle (McConnell 1996). The efforts to fix a bug must, of course, be coordinated by someone in a management role, where it is essential that such a person has a good working relationship with the scientific and infrastructure developers and sufficient gravitas to elicit prompt action. While the task of supporting many compilers and platforms has caused a number of debugging headaches, it has undoubtedly served to improve the quality of the resulting model code. The recent advent of a mature open source Fortran compiler—gfortran[6]—will go a long way to aiding the portability of ESM codes.

In addition to source code, a comprehensive build system can also support other project initiatives. For example, a *make doc* rule in the makefile can be used to invoke an automatic documentation creation tool such as Doxygen (described above). A *tags* target can be used to invoke the Exuberant Ctags[7] tool, to aid source code ("Ctags generates an index file of language objects found in source files that allows these items to be quickly and easily located by a text editor or other utility. A tag signifies a language object for which an index entry is available." As an example, given a tag index, an appropriately equipped text editor can jump a developer from a subroutine call directly to the definition of that routine, even if that definition is in another source code file or directory.) Further variables and conditionals can be added to a makefile

---

[4] http://www.cmake.org

[5] http://www.cdash.org

[6] http://gcc.gnu.org/fortran

[7] http://ctags.sourceforge.net

to activate particular compiler flags of benefit when profiling or debugging the source code, e.g. *make BUILD=DEBUG*.

An ESM will typically require a number of tools to be present upon a system it is to be run on: a Fortran compiler, the make utility, NetCDF libraries etc. A software infrastructure developer will have to weigh the cons as well as the pros of this 'footprint of dependencies'. For example, mixed-language programming can offer a number of benefits: number crunching duties can be efficiently carried out by a library of Fortran (or C/C++) routines, whereas user-interface or 'glue' can be better written in higher-level languages, such as Python for example. However, introducing additional programming languages adds an extra burden to the source code developers and to the users and the systems upon which it is hoped that the resulting creation will compile and run.

A class of model dependency which is particularly fraught are those which must be compiled by the user, but which fall outside of the build system. An example of such a dependency are the excellent NetCDF libraries helpfully provided by Unidata.[8] These are a significant aid to the goal of a standardised file format for model inputs and outputs. In our experience a great many model users have experienced difficulties in either finding, building or linking to these libraries on their computer systems. Helping to overcome these glitches from afar, without any access to the computer systems in question, has required the acquisition of very particular debugging skills and expertise on the part of the software infrastructure developers. A common issue in this regard has been the 'name mangling' process performed when creating object code and finding a solution has required an exposition on the nuances of library file creation. The 'footprint' of the GENIE framework includes Fortran and C/C++ compilers, Python, and XSLT engine such as xsltproc, make, the bash shell and the NetCDF libraries, together with Matlab and the proprietary OPTIONS toolbox for model tuning (see Vol. 1 for a more detailed description).

7.4 Running and Monitoring

An important precursor to running any experiment is to thoroughly test and validate any model installation. Analogously, it would be meaningless to use the output from a mass spectrometer without first calibrating the machine. In the GENIE framework, it is an easy matter to designate certain configuration files as test cases, since their content is assured and tracked by virtue of being held in the version control repository. There are at least two categories of test which may be defined. One set of tests may be used to determine whether the installation of the model code on a different computer system has been successful. That is to validate a *port* of the model. Another set of tests may be used to check whether the output of the model has changed in any way, following some modification of the code. A *regression* test. These two sets may, of course, intersect. For non-chaotic ESM formulations, such

---

[8] http://www.unidata.ucar.edu/software/netcdf

as one using an energy-moisture balance atmospheric model, the small differences between model runs introduced at the magnitude of machine rounding by changes in compiler, microchip architecture etc. are not magnified and so model outputs can be checked against reference copies stored in the repository. The picture is not so neat for ESMs which contain dynamic general circulation component models, however, since the embodied equations are highly sensitive to initial conditions and so small differences between runs quickly lead to a large divergence in model outputs. When output differences arise due to running the model on two different computers, it may be possible to determine an acceptable threshold upon the rate of divergence of model runs in order to validate the port (Rosinski and Williamson, 1997). Following a similar approach for the task of regression testing within the GENIE framework, it was not possible to find a threshold which could distinguish between a difference introduced by a tiny perturbation to the input atmospheric data and that created by a (deliberately inserted) bug. Thus for regression testing, bitwise comparable outputs are demanded and so platform specific reference files must be used. These are arguably too numerous to be held under version control and so fall outside of any centralised management.

Model tuning is a workflow element common to all ESMs (and especially to the fast, highly parametrised models currently comprising the GENIE framework). The essence of a good tuning exercise is to thoroughly and efficiently examine model behaviour across the multi-dimensional space of possible parameter settings. This can, however, be laborious work. Performing the search in a semi-automated way is significantly aided by the use of XML configuration files which can be parsed (and hence programmatically perturbed) and also written by a number of mature software tools. A vector of parameter settings is adjudged to be good through comparison of model output to reference data (either direct observations or proxy values for climate system characteristics). Since climate has many aspects, this comparison (and hence the tuning exercise) has multiple objectives. Also, since we know that both model predictions and reference data contain errors, it is wise to select a cohort of the best parameterisations, rather to rely upon a single (but likely to be biased) 'winner'.

Running large ensemble experiments, such as a tuning exercise, can require a great deal of computation. The National Grid Service[9] coordinates access to a large pool of compute resource in the UK. (The heterogeneous nature of this resource means that non-chaotic ESMs can be run, but can be problematic for the combination and comparison from models containing dynamic components.) In order to manage the workflows for such large ensembles, the GENIE framework has benefited from the middleware (i.e. intermediary software) developed by the GEODISE project.[10] The Grid does not just offer computational resource, but also the ability to archive model results in a shared repository which may be accessed from distributed locations using standard web protocols. The association of metadata (again in XML format) with model output is key to efficient post-processing of the results held in such an archive. The GENIELab software developed as part of the GENIE project

---

[9]   http://www.ngs.ac.uk

[10]   http://www.geodise.org

coordinates just such an archival process, as well as providing a convenient GUI for monitoring ensemble experiments underway on the Grid. The use of a GUI for running and monitoring the model as well as the importance of advanced features such as data mining, and multi-objective schedulers as described above were highlighted in Chap. 6.

Another direction for running and monitoring a model which is important for policy making is in conjunction with an economic model, as described in Chap. 8. The use of XML configuration files again eases this connection, as does a clear separation of parameters specific to component models from those relevant to the ESM as a whole. Modularity of the framework and standardisation of component model interfaces (as far as is possible) is also key to the flexible coupling required by this enterprise.

7.5 Discussion

Looking to the future, the creation of flexible ESMs capable of being run over many different platforms—as strived for in the GENIE framework—will need to surmount two key challenges. The first is an improved modularity of component models and assurance of their integrity. Establishing well defined and serviceable interfaces between component models is a difficult task, since any partitioning of the Earth System is in some senses an artifice. However, being able to break up the task of developing a new ESM into bite size chunks is essential if we are not to drown in the extreme complexity that large ESMs now embody. No one person has oversight of all the processes contained within today's ESMs. Well designed interfaces will require a collaboration between scientists and software engineers and will considerably simplify the task of component model coupling. Object oriented programming approaches are also likely to be useful in this endeavour.

A workflow element which is essentially missing from all ESMs at this time is the ability to easily test and validate component- and sub-models individually. Remedying this omission will significantly enhance collaborative development efforts as well as user confidence in model outputs. Many mature testing tools already exist and can be brought to bear on the task of determining whether component-models are fit-for-purpose. Examples of these aids include unit testing frameworks (a cornerstone of the Extreme Programming methodology[11]), distributed testing apparatus, like Ctest, and aggregation tools, such as CDash.

The second key challenge is to establish flexibility at the algorithmic level. To date component-models within GENIE have been designed to run efficiently on single-core microprocessors. However, multi-core processors are in the ascendancy and will soon be surpassed by many-core architectures, possibly similar to those exhibited by newly available General Purpose Graphics Processing Units (GPGPUs). In the future modelling frameworks which aspire to be run over a wide variety of platforms

---

[11] http://www.extremeprogramming.org

must contain algorithms which are capable of truly massive parallelism. Today's fastest computers contain hundreds of thousands of processors. As the number of processors available within a single system moves into the millions, the mean time between component failure will shrink and so the chances of losing a processor while performing a massively parallel computation pushes ever higher. Algorithms of the future will need to accommodate this occurrence gracefully and so are likely to employ Monte Carlo methods.

It is clear from the previous chapters that good version control and build systems are essential elements of an ESM workflow. Happily, mature tools to perform these functions are freely and widely available. We can contrast these tools with others in the workflow which are less mature, but have the potential to greatly enhance the workflow overall. Tools for configuration is one such area. While many developers have identified the potential of XML for describing model configurations, a widely adopted *process* for managing configurations has yet to emerge. In a similar vein, the advent of mature monitoring and data mining tools have the potential to hugely enhance a climate scientist's lot, freeing him or her to focus on an improved understanding of Earth system processes, rather than becoming bogged down the details of (re-)submitting jobs and post-processing model output files.

References

Pilato C, Collins-Sussman B, Fitzpatrick B (2008) Version Control with Subversion. http://svnbook.red-bean.com

McConnell S (1996) Rapid development: taming wild software schedules. Microsoft Press, Redmond, p 72

Rosinski JM, Williamson DL (1997) The accumulation of rounding errors and port validation for global atmospheric models. SIAM J Sci Comput 18(2):552–564

Chapter 8
Configuring, Building and Running Models in CIAS

Sudipta Goswami and Rachel Warren

8.1 Introduction

The Community Integrated Assessment System (CIAS) is a type of Integrated Assessment Model (IAM) developed at the University of East Anglia to analyse the linkages between the earths climate system, the impacts of climate change in various sectors, and the global economy. The scientific objectives of IAMs are significantly different from Earth system models (ESMs), and this is reflected in the design philosophy. IAMs link across disciplines, which may not necessarily include climate modelling. Typically ESMs form one component of an IAM. IAMs are designed to answer policy questions and therefore need to be computationally cheap and sufficiently user-friendly to cater for a diverse range of audience. ESMs such as the Met Office Hadley Centre suite of models (Chap. 4) are neither computationally cheap nor capable of reaching out to a diverse audience.

Unlike many IAMs, CIAS was designed to provide powerful tools to understand the robustness of outputs to uncertainties, both within and between models. In particular, CIAS contains several alternative integrated assessment models, called couplings, that can be accessed from a web portal. This allows the user to test the robustness of outputs to the use of different model combinations in answering the same scientific question. It also features user-friendly methods to carry out large ensembles employing techniques such as Latin hypercube sampling (Iman et al. 1981). Also unlike many ESMs, CIAS was developed with stakeholder participation right from the beginning (Warren 2002) and flexibility was a key design requirement.

S. Goswami (✉) · R. Warren
Tyndall Centre, University of East Anglia, Norwich, UK
e-mail: s.goswami@uea.ac.uk

R. Warren
e-mail: r.warren@uea.ac.uk

R. Ford et al., *Earth System Modelling – Volume 5,* SpringerBriefs in Earth
System Sciences, DOI: 10.1007/978-3-642-23932-8_8, © The Author(s) 2012

Hence, CIAS was designed to provide a robust and modular framework to enable multi-institutional collaboration. This implied substantial flexibility in the framework to allow for the different programming languages, operating systems, compilers and run-time environments used at participating institutions. Our design requirements therefore included:

1. Flexibility in choosing couplings (i.e. alternative combinations of linked modules).
2. Flexibility in setting up uncertainty analyses through execution of large ensembles and adjustment of model parameters generally.
3. Relative ease in configuring and executing coupled model runs.
4. Flexibility in allowing linkage of codes written in multiple programming (and scripting) languages and run-time environments (which includes compilers, operating systems and third-party libraries).
5. Flexibility in allowing simultaneous use of multiple hardware resources.
6. Flexibility to run different parts of a model in physically distinct locations, i.e. to work across institutional firewalls.
7. Provide security to control access to not only the entire system but also to modules and results by different users to respect the intellectual property rights of contributing institutions.
8. Provide a central repository to store results.
9. Provide post-processing tools for visualisation and analysis.

8.2 CIAS Framework

The software developed to meet the above requirements is SoftIAM, which relies on a number of readily available and free tools and technologies to provide this flexible framework. This section describes the CIAS framework developed to execute distributed coupled models. The term coupled model refers to a combination of modules which have been contributed by different institutions into a common pool. Modules can, and typically are, individual pieces of software (usually representing some part of the global climate and/or economic system) which have been adapted to work within CIAS. An important point to note here is that the scientific validity of the coupled models is ensured by the contributing scientists and not SoftIAM.

The principal components of the CIAS framework are shown in Fig. 8.1. At the heart of CIAS is the Bespoke Framework Generator (BFG, see Volume 3), which allows flexible composition and deployment of coupled models. BFG generates bespoke source code to couple modules using the framework metadata, which is provided through a series of XML files. SoftIAM is implemented as a thin wrapper around Apache Ant (Sect. 6.7) to automate the build, deploy, configuration and running of a distributed model. SoftIAM calls BFG to create the initialisation and communication routines between modules as well as the deployment rules for distributed execution. As SoftIAM is a command-line tool, a web-portal is used to set-up and configure experiments, display job status and provide results when available.

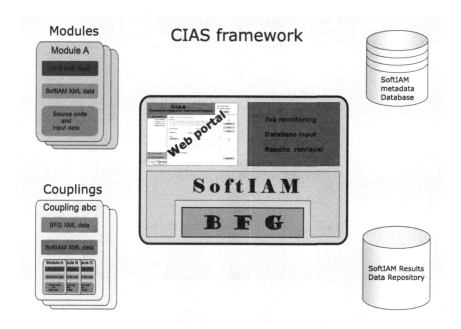

Fig. 8.1 Principal components of the CIAS framework

Additional libraries are used to monitor job status and for database management. The other important components are the modules, couplings and a suitable database to store the metadata for each experiment.

Modules are the adapted codes that are set-up to run sequentially in the system to form a coupled model. Each module is represented by a JAR (Java ARchive) file containing all files (i.e. source code or executable binary of the model, input data and configuration files) required to run the module as well as a configuration XML file. This Extensible Markup Language (XML) file contains information used to auto-generate user interface components in the web portal, build and environment information, and run-time requirements for the module. Some modifications to the module source code are also required for compliance with SoftIAM. Modules can reside in host institutions and be executed locally. The data exchange between modules is controlled entirely through SoftIAM.

Couplings are the scientifically valid combinations of the different modules forming the Integrated Assessment Models. Each coupling is defined using a set of XML files which contains definition and configuration information required by BFG and SoftIAM to define the coupling. The XML schema for a coupling in SoftIAM defines the list of modules that are coupled together, the order in which they are to be executed, how the modules are linked together, i.e. how the outputs of one model are transferred to be the inputs of another module, including any conversion factors e.g. units, the locations where the module codes are stored and executed. Couplings are also stored in the system as JAR files.

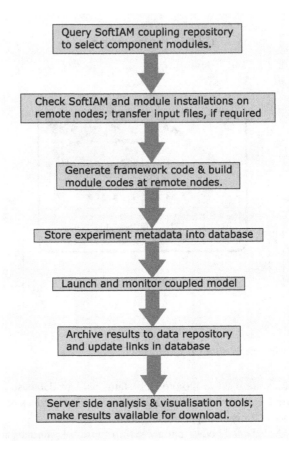

Fig. 8.2 SoftIAM workflow after user-initiated start of an experiment

The CIAS framework incorporates an off-the-shelf database to store metadata related to each experiment. Metadata typically includes the selected coupling name, and its configuration. Currently Hypersonic SQL[1] is used for this purpose as it is free, lightweight and is bundled with the JBoss[2] application server. The latter is used to run the web-portal. The extremely large sizes of results files preclude it from inclusion directly into a database. A data repository, typically the machine hosting the web portal, is used to store the data files with soft links saved within the metadata.

SoftIAM (which includes BFG) must be installed on every machine (called compute nodes) which will run component modules. The web-portal, metadata database and data repository is installed on the principal node. All coupling jar files must be kept on the principal node. Modules need to reside only on machines where they

---

[1] Hypersonic SQL is an open-source database written in Java and packaged with the JBoss application server.

[2] http://www.jboss.org

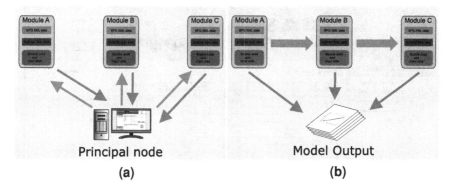

Fig. 8.3 a Control structure for an experiment. The principal node, which hosts the CIAS-portal and SoftIAM, controls the flow of an experiment by sending instructions to remote nodes sequentially. The actual experiment flow is shown in **b**, in which all modules create output data that is collated for archiving by SoftIAM

will be executed. All couplings must be deployed on relevant nodes, which ensure that a valid installation of SoftIAM and the module(s) are present on the appropriate compute nodes. Communication between remote nodes and the principal node is through secure-shell for deploying experiments and retrieving results.

8.3 Experiment Configuration and Execution

User interaction with SoftIAM is mainly through the CIAS portal, which is a web-based application that simplifies the process of specifying and running a coupled model and viewing the results. The web-portal also ensures user access control, as the CIAS system administrator can select which modules the users will have access to. This is an important feature, as some participating institutions need to restrict access to their code until it is peer-reviewed. When the user selects a particular coupling, the user interface for the component modules is dynamically generated from the information stored in the module configuration (XML) file. The CIAS portal allows the users to run any number of experiments using the same or different coupled models, and even large ensembles.

The CIAS workflow (Fig. 8.2) is quite similar to the generic workflow discussed in Fig. 2.1. Once a user selects a coupled model and starts an experiment, SoftIAM checks for a valid installation on all the systems where the model will run. Note that the module source code must be located on the same machine on which the executable for that module is to be built and run, unless module codes are provided as binary libraries or cross-compilation is possible. SoftIAM then transfers all input files to relevant compute nodes, if required (e.g. if provided by the user) and builds all the modules. Prior to starting an experiment, SoftIAM stores all the metadata for an experiment in the database. This ensures that the experiment can be re-created in

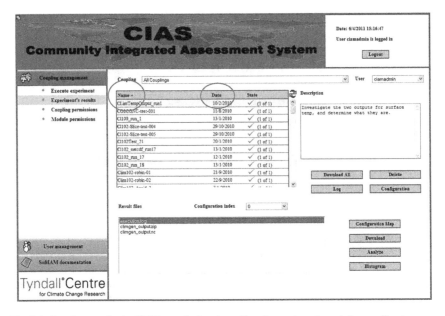

Fig. 8.4 Results page in the CIAS portal, showing a list of experiments sorted according to experiment ID. The list can also be sorted according to the date of execution as well

case of catastrophic hardware failure or network outage. Job monitoring in CIAS is still at a nascent stage. Currently, there is no queueing system or load-balancing in place and the system waits for the return code from a particular SoftIAM command. This is one area, where development effort will need to be concentrated in the future.

During the progress of an experiment, output data from each module is sent back to the principal node when the module completes execution (Fig. 8.3). SoftIAM then passes the data to the next module in the sequence. Of course, only data required by the next module is sent. Communication between modules and SoftIAM is implemented using MPICH2. MPICH2 is a portable implementation of the Message Passing Interface (MPI) and is one of the protocols used to communicate between remote machines in BFG. On completion of an experiment, all the data is moved to the data repository, tagged and the ID entered into the database. Archived data from previous experiments can be extracted from the data repository, using either the experiment identification string or the date, as shown in Fig. 8.4.

8.4 Conclusion and Future Development

The CIAS framework has been designed to run integrated assessment models composed from modules provided by different institutions. The system design requires the ability to compose different model combinations from modules which are distributed across institutions and written in different languages and are run under different run-time environments. For this, the modules are adapted to separate meta-data about the model from the scientific code. This allows the module code to operate in any coupling and allows high level of flexibility in deployment. Users operate the system via a web application called the CIAS Portal which simplifies the setup, running and monitoring of experiments as well as retrieving results. The SoftIAM workflow has been made as generic as possible, allowing the use of well-established and freely available third-party tools and technologies such as Apache Ant, secure shell, MPICH2, JBoss AS, XML and Java.

CIAS has evolved, and will continue to evolve, according to the needs of stake-holders and the scientific community. Typically, this has involved the incorporation of new modules, the uploading of new datafiles, the support of more complex model configurations, and more complex data processing and visualisation tools. A number of areas have been specifically identified for further development in CIAS. These include the addition of new modules to create new model combinations which will improve the study of the robustness of outputs associated with any particular emission scenario. It is also hoped that incorporation of a load-balancing and queueing system will vastly improve the scalability of the system. The archival of results in the current setup precludes any method of (meta)data mining for a search & query interface. It is hoped that changes to the underlying connectivity with the database will allow incorporation of an interface that will allow users to search experiments on the basis of individual module parameter values.

References

Iman R, Helton J, Campbell J (1981) An approach to sensitivity analysis of computer models, Part 1. Introduction, input variable selection and preliminary variable assessment. J Qual Technol 13(3):174–183

Warren R (2002) A blueprint for integrated assessment. Technical Report No. 1, Tyndall Centre, http://www.tyndall.ac.uk

Chapter 9
Summary and Conclusions

Rupert Ford and Graham Riley

This book has described the workflow of constructing and configuring (coupled) Earth system models (ESMs), building the appropriate executable or executables from the relevant mix of source code and libraries, and running and monitoring the resultant job, or jobs.

In Chap. 2, Balaji and Langenhorst set the scene with an overview of the whole workflow. They explained how a workflow is essentially a description of what needs to be done which therefore allows for reproducibility of experiments and acts as a record of how data was produced. Their view was that Earth system modelling is in the process of transitioning from being a specialist "heroic" task to being a more general purpose workflow process. They illustrated the current state of the art in workflow for ESM's using FRE and went on to suggest that ESM workflows would benefit from standardisation which would help realise the vision of a simple to use ESM system.

Balaji and Langenhorst's observation leads us nicely on to the next chapter (Chap. 3) where Turuncoglu presented the application of a general purpose scientific workflow engine (Kepler) to ESM. He illustrated the issues with a motivating example in which two instances of CCSM4 with differing resolutions were run on different remote resources. He demonstrated how Kepler could be relatively easily specialised, using ESM-specific interface code (termed actors in Kepler), to support the particular example.

The subsequent three chapters then looked at three separate elements of the workflow in more detail, from the perspective of the major modelling centres. These

R. Ford (✉) · G. Riley
School of Computer Science, The University of Manchester,
Oxford Road, Manchester, M13 9PL, UK
e-mail: rupert@manchester.ac.uk

G. Riley
e-mail: graham.riley@manchester.ac.uk

R. Ford et al., *Earth System Modelling – Volume 5,* SpringerBriefs in Earth System Sciences, DOI: 10.1007/978-3-642-23932-8_9, © The Author(s) 2012

workflow elements were *configuration management, building software* and *running and monitoring*.

In Chap. 4, Carter and Matthews provided insight into the issues that a large ESM organisation must face when developing and constructing ESM models. They identified and discussed a number of distinct roles within such an organisation and outlined the tools currently in use at the Met Office to support these roles (FCM, Subversion and Trac). They mentioned reproducibility as an important issue in configuration management (which resonates with the benefits of a formal workflow as described by Balaji and Langenhorst in Chap. 2) and pointed out that there is a trend towards increased collaboration between different institutions which will present challenges for configuration management.

In Chap. 5, Legutke discussed the requirements of an ESM build system identifying a number of key issues including build times and portability of software between different architectures. She introduced and discussed techniques and tools that are used by ESM centres to help meet their requirements including Make, SCE and FCM and illustrated the issues using SCE as a use case.

Lastly, in Chap. 6, Larsson introduced job monitoring and management. He introduced a number of tools that can be used for job monitoring and management and presented a detailed use case with SMS. Echoing the view by Balaji and Langenhorst that Earth system modelling is transitioning towards more general purpose use he concluded by suggesting that future monitoring and management requirements would be for jobs running on remote cloud or Grid systems and allowing submission and monitoring from any type of networked device at any time of the day.

Up to this point the book had primarily concentrated on the issues faced by the major modelling centres. The final two chapters of this book expanded the discussion to two related areas, long-term and paleo-climate modelling and Integrated Assessment Modelling. In each case the chapters compared and contrasted their requirements and solutions with those from the major modelling centres.

Chapter 7 introduced the GENIE community model which is designed for long-term and paleo-climate modelling. Many of the issues and solutions employed in this model are the same as those described in earlier chapters. However there are also interesting differences. For example in the GENIE community model, portability and performance across different architectures and compilers is considered to be very important and this is reflected in their approach. This contrasts with the major modelling centres who typically run their models on a single dedicated resource.

Finally, Chap. 8 introduced the CIAS Integrated Assessment Modelling system. Again, a number of the issues and solutions employed in this model are the same as those described in earlier chapters. However a notable difference is that Integrated Assessment demands a much greater focus on interoperability between components, as models are written in different institutions and in different computer languages. Further, institutions may place restrictions on where models may be run, requiring support for heterogeneous architectures. Performance is typically less important in IAM as models tend to be relatively computationally inexpensive. Another interesting difference is in the expectations of the requirements of users. The CIAS portal constrains access to pre-defined couplings and choices of parameters to change, whereas Earth system modelling interfaces tend to also offer users the option of modifying couplings (and the underlying scientific code if they so wish).

In each of the previous chapters the authors have provided an indication of the state of software infrastructure associated with their chapter and how it might evolve in the future. We now relate these chapters to two significant trends in ESM, (1) the increase in code sharing amongst institutions and (2) the transition of ESM's from specialist to more mainstream use.

As Earth system models become increasingly complex there has been a gradual increase in the sharing of codes between institutions (as institutions are no longer able to be master of all the different ESM components). Of course this approach is already the norm for Integrated Assessment and systems with a more community oriented approach, such as GENIE.

The trend requires a much greater use of configuration management and version control tools such as Subversion, Trac and FCM, as discussed in Chap. 4, and the use of such tools in Earth system modelling is expected to increase and also be better integrated with emerging workflow management systems.

Further, shared codes may be used in very different circumstances. A researcher in a University may want to run a land surface model in single point mode on their local machine, a major ESM modelling centre may want to integrate that same code with their ESM in order to perform a long running climate change simulation, and a policy maker may want to answer a question on the land-use implications of a certain economic policy, which could result in this same code (as part of an Integrated Assessment simulation) being run on remote resources, such as a cloud, via a query to a portal.

Software infrastructure must be flexible enough to support these different ways of working and not restrict working practices (a point made in Chap. 2 regarding standard workflows). For example, infrastructure should allow both local and remote access to resources, as discussed in Chaps. 3 and 6.

It is unlikely that a single suite of tools will emerge to satisfy all of the circumstances mentioned above. Indeed, in some cases it is considered good practice to minimise dependencies on external software, see Chap. 7. It is more likely that a small suite of tools will co-exist, as is already the case with the use of Ant and Make at the low level and FCM and SCE at a higher level, in the building of models (see Chap. 5) with users choosing the ones that are most appropriate for their particular circumstance. The emergence of underlying standards, such as formal workflow descriptions (see Chaps. 2 and 3) is also expected to help here by improving tool interoperability.

The second trend, identified by Balaji and Langenhorst in Chap. 2, and succinctly put as migrating from the heroic to the mundane, implies an improved ability to support a wide spectrum of users from those with very little knowledge of the underlying infrastructure, or even the underlying science, to the domain specialist. In the extreme case some users may simply want to ask for certain results and will not want to know whether the results were generated by a model as a consequence of the request or the data simply returned from a database.

Interestingly IAM's (see Chap. 8) are already attempting to reach out to the less sophisticated user. Their approach is to provide portals which expose the level of detail that their target audience expects. So, for example, infrastructure choices and the ability to couple models together are typically hidden. The layering of views with different amounts of sophistication onto the same underlying infrastructure seems to be the natural route to take as ESM's transition to more mainstream use.

Glossary

ANT	Another Neat Tool
API	Application Programming Interface
AR5	IPCC Assessment Report 5
BFG	Bespoke Framework Generator
CCSM	Community Climate System Model
CDO	Climate Data Operator
CDP	Command and Display Program
CIAS	Community Integrated Assessment System
CLI	Command Language Interpreter
CM	Configuration Management
CPU	Central Processing Unit
DKRZ	Deutsches Klimarechenzentrum, the German Climate Computing Centre
ECMWF	European Centre for Medium-Range Weather Forecasts
ESM	Earth System Model
ESMF	Earth System Modelling Framework
FCM	Flexible Configuration Management System
FLUME	Flexible Unified Model Environment
FMS	Flexible Modeling System
FRE	FMS Runetime Environment
GCM	Global or General Circulation Model
GENIE	Grid ENabled Integrated Earth system model
GEODISE	Grid Enabled Optimisation and Design Search for Engineering
GFDL	Geophysical Fluid Dynamics Laboratory
GNU	Project to develop a Unix-like operating system sponsored by the Free Software Foundation(http://www.fsf.org)
GPGPU	General Purpose Graphics Processing Units
GPL	GNU General Public License free software license

GUI	Graphical User interface
HPC	High Performance Computing
HTTP	Hypertext Transfer Protocol
IAM	Integrated Assessment Model
IMDI	Integrating Model and Data Infrastructure at DKRZ
IP	Internet Protocol
IPCC	Intergovernmental Panel on Climate Change
IT	Information Technology
JAR	Java Archive
JULES	Joint UK Land Environment Simulator
LAN	Local Area Network
MCT	Model Coupling Toolkit
METAFOR	Common METAdata FOR Climate Digital Repositories
MPI	Message Passing Interface
MPI – M	Max-Planck-Institut für Meteorologie
MPICH2	Message Passing Interface Chameleon Version 2
NAG	Numerical Algorithms Group
NCL	NCAR Command Language
NEMO	Nucleus for European Modelling of the Ocean
NGS	National Grid Service
NWP	Numerical Weather Prediction
OASIS	Ocean Atmosphere Sea-Ice Soil
OPeNDAP	Open-source Project for a Network Data Access Protocol
OS	Operating System
OTP	One-Time Password
OWL	Web Ontology Language
POSIX	Portable Operating System Interface for Unix
PRISM	Partnership for Research Infrastructures in Earth System modelling
ROMS	Regional Ocean Modeling System
RPC	Remote Procedure Call
RSL	Resource Specification Language
SCE	Standard Compile Environment
SMS	Supervisor Monitor Scheduler
SPARQL	SPARQL Protocol and RDF Query Language
SQL	Structured Query Language
SSH	Secure Shell environment
TCP/IP	Transfer Control Protocol / Internet Protocol
TRAC	Tracking system for software development projects
UEA	University of East Anglia
UKCA	UK Community Atmospheric Chemistry-Aerosol global model
UM	Unified Model (from the Met Office in the U.K.)
UMUI	Unified Model User Interface

VCS	Version Control System
VMS	Virtual Memory System
VPN	Virtual Private Network
WRF	Weather Research and Forecasting Model
XML	Extensible Markup Language
XSLT	Extensible Stylesheet Language Translations

Index